Culin E. Von

The Art of Incubation and Brooding

Culin E. Von

The Art of Incubation and Brooding

ISBN/EAN: 9783337380335

Printed in Europe, USA, Canada, Australia, Japan

Cover: Foto ©berggeist007 / pixelio.de

More available books at **www.hansebooks.com**

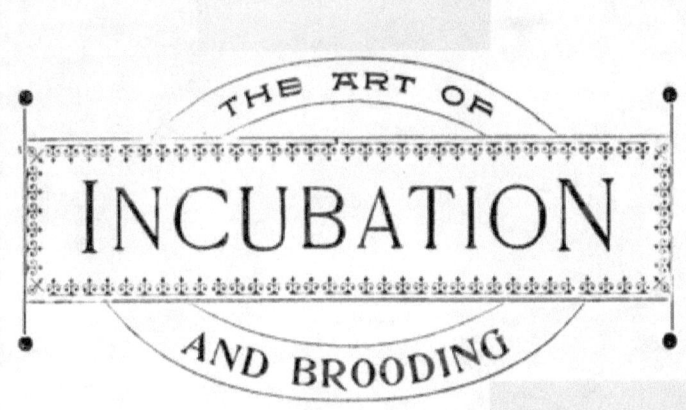

THE ART OF INCUBATION AND BROODING

A GUIDE TO

Profitable Poultry Raising,

BY

E. & C. VON CULIN.

PRICE, ONE DOLLAR.

DELAWARE CITY, DEL.
E. & C. VON CULIN.
1894.

E. VON CULIN.

C. VON CULIN.

PREFACE.

This book is written to aid and inform beginners who know little or nothing of artificial incubation and brooding; to assist those who have learned something about it and wish to know more; and to supply a handy reference for those who know it all.

As originally written, it would have made a volume three times as large as the present one, but the people of to-day want everything boiled down, concentrated, concise, convenient. A triple extract contains all the perfume and vital qualities of three times its bulk of plain tincture; so we have rewritten it, carefully eliminating all superfluous matter, and placing the gist of it before our readers in a compact form, convenient for the desk and not too cumbersome for a good-sized pocket. Instead of using a thick, porous paper, which would increase the bulk fourfold, we have chosen a heavy, superior plate paper and new type, believing that our readers would prefer elegance to unnecessary bulk.

INCUBATION IN EGYPT.

Artificial incubation is almost as "old as the hills." It was known and practised in ancient Egypt, and is to-day an important industry of that interesting country.

While no monumental picture of an incubator has been discovered, the authorities are a unit in the belief that the Egyptian hatching houses of the present time are substantially the same as those of prehistoric Egypt.

Diodorus of Sicily speaks of it as an art that had been in use a long period before his time. Pliny says nearly the same. The Roman Emperor Hadrian found it generally practised in Egypt, and makes special mention of it in his description of the usages and customs of that country.

A French missionary, who traveled in Egypt in 1737, says: "I found there were about four hundred chicken-ovens, each one furnishing about two hundred and forty thousand fowls, making about one hundred millions produced each year from this source alone. In selling them, they do not count them, but measure them by the bushel, like grain. Though there are always some smothered, this process saves the trouble of separating them according to quality and size. In attempting to ascertain the origin of this practice of artificial incubation and to explain its success, two facts should be noticed; first, that it was exceedingly useful to multiply the amount of food as healthy as that furnished by the flesh of birds; and second, without some such process fowls of all kinds would have become very scarce, for the reason that the heat is so great in the laying season that the pullets abandon their eggs for the society of the cocks. Finally, geese, ducks and other fowls are also multiplied by incubation."

The Egyptians of to-day are extensive raisers of all kinds of poultry, and as hens do not sit well in that or any other hot country, most of the birds are hatched in artificial hatcheries or incubating houses, which, on account of their necessarily large size and consequent expense of building and management, are not built upon the farms of the poultry raisers, but are owned principally by Copts, who make a business of hatching for the farmers and villagers on shares or for a stated price, the eggs being carried to them by the poultry keepers, who receive the birds, or their share of them, when hatched and ready to remove.

Notwithstanding the fact that they cling to their primitive style of incubation, the Egyptians are among the most successful in artificial hatching, and it is worth while to note that they use no hot water tanks in their hatching rooms. Hot air suits them better because it is easier to control than hot water, and more economical—important items in a hot country and where fuel is high.

Though slow to adopt improvements, it is probable that they will ultimately make use of the portable incubators which require no night attendant, and with which each farmer may do his own hatching at home. More than ten years ago a French company in Egypt were using portable incubators in hatching ostriches at their ostrich park at Matareeyeh, which lies ten kilometres northwest of Cairo and one kilometre south of the ruins of Heliopolis, not very far from the old railroad which runs direct from Cairo to Suez.

EGYPTIAN INCUBATING HOUSE.

The "mahmal" or incubating house is built of sun-dried bricks and contains from eight to twenty-four ovens. On each side of a passage is a row of ovens and fire-places. The ordinary size of the ovens is 10 feet long, 8 feet wide and 6 feet high. The fire-places are above the ovens, and are the same length and width as the ovens, but not so high. There are doorways to each oven, large enough for a man to enter, and a small opening between the ovens and the fire-places. Besides this there is an opening connecting all the fire-places. The latter have places for the smoke to escape, and there are also chimney holes in the roof of the passage but they are seldom opened. The eggs are placed in the ovens upon mats or in chopped straw, in tiers one above the other, usually not more than three high. The fuel used is "gellah," made of the dung of animals mixed with chopped straw and moistened to form flat, round cakes, which are sun-dried before they are used. Only half of the number of ovens are used the first ten days, and the fires are lighted upon the fire-places above them only. On the eleventh day these fires are extinguished and other fires are lighted on the remaining unused fireplaces. Then some of the eggs are removed from the first set of ovens to the fire-places above, which are, of course, still heated, though the fire has been removed. When the first eggs are hatched and the second half hatched, fresh eggs are placed in the position made vacant by those first hatched. This rotation continues until the hatching season is over. The chicks are kept in a warm room for two days and then delivered to the various parties for whom they are hatched.

Sir J. Gardner Wilkinson, in his "Popular Account of the Ancient Egyptians, 1854," describes the ovens and the process as follows: "The modern process, like that of ancient times, is this: they have ovens expressly built for the purpose; and persons are sent round to the villages to collect the eggs from the peasants, which, being given to the rearers, are all placed on mats, strewn with bran, in a room about 11 feet square, with a flat roof and about 4 feet high, over which is another chamber of the same size, with a vaulted roof and about 9 feet high; a small aperture in the centre of the vault (at f), admitting light during the warm weather, and another (e) of larger diameter, immediately below, communicating with the oven through its ceiling. By this also the man descends to observe the eggs; but in the cold season both are closed, and a lamp is kept burning within; another entrance at the front part of the oven, or lower room, being then used for the same purpose and shut immediately on his quitting it. By way of distinction I call the vaulted (A) the upper room and the lower one (B), the oven. In the former are two fires in the troughs $a\,b$, and $c\,d$, which, based with earthen slabs, three-quarters of an inch thick, reach from one side to the other against the front and back walls. These fires are lighted twice a day; the first dies away about midday, and the second, lighted at 3 P. M., lasts until 8 o'clock. In the oven the eggs are placed on mats strewn with bran, in two lines corresponding to, and immediately below, the fires $a\,b$, and $c\,d$, where they remain half a day. They are then removed to $a\,c$, and $b\,d$; and others (from two heaps in centre), are arranged at $a\,b$, and $c\,d$, in their stead, and so on till all have taken their equal share of the warmest positions, to which each set returns again and again, in regular succession, till the expiration of six days.

"They are then held up, one by one, towards a strong light; and if the eggs appear clear, and of an uniform color, it is evident they have not succeeded; but if they show an opaque substance within, or the appearance of different shades, the chickens are already formed, and they are returned to the oven for four more days, their positions being changed as before. At the expiration of the four days they are removed to another oven, over which, however, there are no fires. Here they lie for five days in one heap, the aperture (e, f) and the door (g) being closed with tow to exclude the air; after which they are placed separately about one or two inches apart, over the whole surface of the mats, which are sprinkled with a little bran. They are at this time continually turned and shifted from one part of the mats to another, during six or seven days, all air being carefully excluded, and are constantly examined by one of the rearers, who applies each singly to his upper eyelid. Those which are cold prove the chicken to be dead, but warmth greater than the human skin is the favorable sign of their success.

"At length the chicken, breaking its egg, gradually comes forth; and it is not a little curious to see some half exposed and half covered by the shell; while they chirp in their confinement, which they show the greatest eagerness to quit.

"The total number of days is generally twenty-one, but some eggs with a thin shell remain only eighteen. The average of those that succeed is two-thirds, which are returned by the rearers to the proprietors, who restore to the peasants one-half of the chickens; the other being kept as payment for their expenses.

"The size of the building depends, of course, on the means or speculation of the proprietors; but the general plan is usually the same; being a series of eight or ten ovens and upper rooms, on either side of a passage about 100 feet by 15, and 12 in height. The thermometer in any part is not less than 86° or 88° Fahr.; but the average heat in the ovens does not reach the temperature of fowls, which is 104°.

"Excessive cold or heat are equally prejudicial to this process; and the only season of the year at which they succeed is from the 15th of Imsheer (23d of February) to the 15th of Baramoodeh (24th of April), beyond which time they can scarcely reckon upon more than two or three in a hundred."

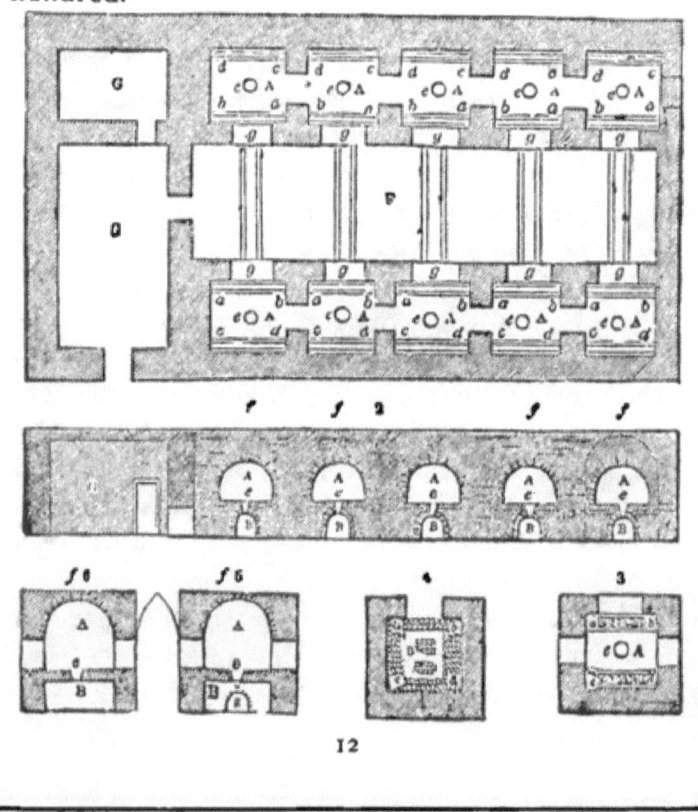

Fig. 1, plan of hatching house, A A A A, plan of the upper rooms or fireplaces; F, passage between ovens; *g g g g*, doors; *e e e e*, opening between ovens and fireplaces; G G, rooms for attendants, fuel, etc. Fig. 2, section of hatching house; A, upper rooms or fireplaces; B, lower rooms or egg ovens. Fig. 3, plan of upper rooms or fireplaces; fire placed at *a, b, c, d.* Fig. 4, lower rooms or ovens, showing eggs in place. Figs. 5 and 6, sections from back and front of upper and lower rooms.

A GOOD INCUBATOR.

To succeed in raising poultry, either for broilers or for egg production, it is necessary to have a *good* incubator. A tolerably good or ordinary one will not do; you should have the very best you can find, regardless of cost, because the first cost cuts a very small figure when compared with the losses which always follow the purchase of a "Cheap John" incubator (?). Of course, you know there are various kinds of cheap affairs misnamed incubators.

The great demand for incubators and brooders has tempted sash manufacturers, makers of show cases and others, to get out various boxes, cases, tanks and barrels, with various attachments, and call them incubators or hatchers. Some buy a lot of almost expired patents, and boom the new machine on the reputation of the old one, to which the patents originally applied, while the new machine possesses none of the good points of the old one, which to build would cost considerably more than the new one is sold for.

Many of this class never had any merit, and went out of the market, but new ones bob up along the line, have their day of deceit and disappear. Watch for them.

The wonderful improvements in mechanism in the last few years have made it possible to procure a first-class machine in almost any line, and incubators and brooders have moved in the front rank of progress. There is no difficulty in getting a good incubator, provided you are careful to avoid bad ones and are willing to pay a fair price for the best.

Do not be deceived by a similarity in names of incubators or hatchers. It sometimes happens that a sharper, with plenty of cash to advertise and push a business will drive a rattling trade in a good-looking inferior machine for several years, making thousands of sales and hosts of unhappy purchasers. Then, when complaints pile up like blocks of ice in a gorge, and respectable papers begin to question the wisdom of allowing the advertisement to remain, and the business begins to decline, he gets out another cheap trap and gives it another name, *similar* to that of a first-class hatcher, which has an established reputation. Sharps or sharks of this kind sometimes hatch out a new firm name with the new machine, and the people make a rush for the cheap affair, many of them confusing or confounding it with the meritorious machine, whose name it approaches as closely as possible, without possessing a single feature, except the name, which could be called even a weak imitation. These tricksters disappoint their customers and injure the reputation of a good machine. Look out for them.

HOW TO CHOOSE AN INCUBATOR.

We are not going to tell you which make of incubator to buy; but we are going to try to show you how the various kinds of incubators are constructed, their several methods of operation, and the good and bad points of each method, with our reasons for calling them good or bad, as the case may be. Having used nearly all the makes now on the market, and many that have gone out of the market, we should be able to do so very fairly. Then we shall leave you to judge for yourself. This you can do intelligently if you will carefully examine each machine, either by actual sight of it, or by the illustrations which their manufacturers send out to inquirers.

First look at the machine, or a picture of it, read the description which accompanies it, and be sure that you understand how it works. If the principles are clear to you, then consider whether or not the application of those principles as shown, will produce the results essential to successful incubation—if, in your opinion, they will accomplish all that the manufacturer claims for his machine.

If the construction and principle are correct, the maker can have no good reason for failing to show them plainly to his prospective customers.

When a manufacturer fails to show and explain the interior work and construction of an incubator, you will be pretty safe in your conclusion that either he has nothing good to show, or he has something bad to conceal. These are enlightened days, and the average man or woman readily understands the artist's lines when accompanied by simple explanations.

Do not be deceived by handsome appearance, big claims, or miraculous testimonials. Beauty is no objection, if it is not made a substitute for utility. Good testimonials are highly valuable, but look into them—write to a few of the persons who give them.

Do not be deceived by low prices. The best article cannot be made, much less sold, at the price asked for the poor or bad one. A house is a house, yet no one expects to get as good a house for $1,500, as he can build for $3,000. No one will say that one house is just as good as another, regardless of plan, material or cost.

DON'T MAKE A FAILURE

For the sake of a few dollars on the start. So many people say, "Well, I will buy a cheap incubator and try the hatching business, and if I succeed with it, I can then buy a better one." This is false economy. It is like buying a poor horse to go a long journey. The horse fails to carry you to your destination, and when he gives out on the way, you must either buy a better horse, or walk.

Nine out of every ten failures in poultry raising are due to false economy at the start. No farmer who knows anything about harvesting would use a scythe in preference to a modern self binding reaper because the scythe is cheaper. Nor will he buy any but the *very best* reaper, even if it does cost a little more than some others. It is true economy to have the best and to start right.

THE BEST SIZE INCUBATOR.

Many beginners are undecided as to what sized incubator to get. If we wanted a capacity of 300 eggs would get three incubators of 100 eggs capacity each; if 600 capacity, three of 200 eggs each; if 750, three of 250 each; if 1200 capacity, three of 400 each; if 1800 capacity, three of 600 eggs each. This is much better than getting one large incubator for all the eggs. It costs more for the several smaller machines than for one large one for all the eggs, but the advantages are: You can have fresher eggs for each incubator, you can sort the eggs if you have large quantities, and select those with shells of same kind and thickness for each incubator; you can place duck, turkey or goose eggs in separate machines, or use a different machine for each variety of hens' eggs. You can keep a record of each kind and quality; you will learn more about the amount of moisture for each class of eggs, and will soon become able to hatch all kinds of eggs equally well. If you make a mistake you will discover it more easily and can rectify it more readily; the result of a mistake or an accident will not be as expensive, and you will have a better chance to retrieve any loss which you may sustain through accident, carelessness or neglect of rules in hatching, for it would hardly be likely to affect but one machine, and as that one would only contain one-third of your full quota of eggs, you would have the other two-thirds left even if all in one machine were ruined, and you would not be apt to repeat the performance (or non-performance) with either of the other two incubators.

HOT AIR OR HOT WATER?

A question which is sure to confront the beginner (and the old poultryman who determines to use the incubator instead of hens) is: Which is preferable, a hot air or hot water incubator?

No one can answer that question better than yourself, if you will just look into it, and make good use of your natural intelligence. Hear argument on both sides, but examine the evidence, look at the apparatus, there is nothing hidden, mysterious, or complicated about it. The principles are as simple as addition and subtraction. You do not have to ask a professor of mathematics whether or not two and two make four, or which is the greater number, one or two? Neither will you have to depend upon any man's advice in this matter, if you have ordinary intelligence and the self confidence which every man must have to succeed in the poultry (or any other) business.

The reason that so many intelligent persons go wrong in this matter is that they do not stop to think.

There is so much nonsense written and published, that the man who will not do his own thinking and reasoning will almost certainly get bewildered. For instance a writer in a prominent poultry paper recently wrote advocating hot water incubators "because hot water heat is *moist*," and suggested that it was the duty of those possessing "knowledge" to give it to the fraternity through the columns of the poultry papers (and the editor published the nonsense without a word of comment).

In another poultry journal an advertiser of a hot water incubator, says: "Hot air is necessarily foul air. Hot water is next to hot blood, the hen's life giver to eggs."

Did you ever hear such philosophy?

We do not need a sage to tell us that—"*hot water is moist*"—but we have yet to discover an incubator in which the eggs are placed in the water. Neither do we know of one in which the hot water tank is open to the egg chamber; nor would a single chick hatch in an incubator with the heater tank so opened.

Nor have we any knowledge of blood sweating hens.

The heating tanks of all hot water incubators are both air-tight and water-tight as far as they connect with the egg chamber, and if they never leaked, not a single drop of moisture would ever get into the egg chamber from that source; and except in those machines having a top opening or tube from the hot water tank, in which a float or other device is used to operate on a regulator lever or valve, by expansion of water in said tank, the water in the tank would never evaporate or grow less, but would suffice for running the machine through one or a dozen hatches, except it were necessary (as is frequently the case) to draw off a quantity when overheated and to add cold water to hasten the reduction of temperature.

But as far as supplying moisture to the eggs—not a drop is supplied by the hot water tank. *There is no moisture from the hot water!* Now where is the "moist heat?"

How do all hot water incubators, as well as all hot air incubators, get their moisture?

From open pans above or below the eggs.

If moisture oozed from the hot water tank there would be no necessity for moisture pans.

Now let us look at the hot air incubator. It has a heater over the eggs, just as the hot water machine has. You can call it a tank or a reservoir, and it is perfectly air-tight in relation to the egg chamber. The egg chamber is heated by radiation from the lower surface of this tank or reservoir, and it gives just as much moisture to the eggs as does the heat which radiates from the lower surface of the hot water tank—which is *none whatever!* Neither can it give any fumes, because neither gas nor water can penetrate the metal radiator which has no opening whatever into the egg chamber. The heat or hot air does not pass from the lamps to the egg chamber any more than it does in the hot water incubator.

The moisture is supplied from open pans above or below the eggs, just the same as it is in the hot water machine.

Now in the face of these *facts* is it not ridiculous, not to say insulting to the intelligence of poultrymen, to assert that the heat from a hot water tank is moist heat.

We will go farther and say that the best hot water incubators use moisture pans with from four to ten times the area of evaporating surface used in the modern hot air incubator. If the heat from the hot water tank is moist, why use more water in the egg chamber than is used in the hot air machine?

Now, let us see what logic there is in the assertion that "hot air is necessarily foul air," when applied to an incubator.

The egg chamber of a modern hot air incubator contains atmospheric air which is drawn into it from the room in which the machine stands, through ventilators having no connection whatever with the heat reservoir, and the heat reservoir has no opening whatever into the egg chamber. The bottom of the heat reservoir is sheet metal and forms the top or ceiling of the egg chamber.

When the lamps are lighted the latent heat in the oil is gradually evolved, and passes up through the fire-proof conductors into the reservoir. When a sufficient quantity of this heat or *caloric* accumulates in the reservoir—being supplied faster than it can pass out through the draft tubes of the reservoir, which is open to the outer air *above* it, it begins and continues to radiate from the metal bottom of the reservoir and is diffused throughout the egg chamber. Nothing passes through this metal radiator but the heat; no air, no gas, no odor—simply the mysterious agent which chemistry has named *caloric*.

Suppose we fill the heat reservoir with water, coffee, milk or wine, and heat the liquid say, to 150°, what follows? Why the heat (*caloric*) radiates from the metal bottom of the reservoir and is diffused through the egg chamber, just as it does when the reservoir contains air, and this radiated caloric or heat mingles with the air in the egg chamber and is absorbed by the eggs until they and the air immediately surrounding them are heated to 102°, 103° or 104°, according to the desire of the operator.

The air in the egg chamber thus becomes warm air or "hot air," and if there is good ventilation in the egg chamber, the hot air is not foul air.

Now, what becomes of the assertion that "hot air is necessarily foul air."

When it comes to the question of hot water heaters or hot air furnaces for heating houses, an entirely different principle is involved. Although you get no moisture at all from hot water pipes, you get pure heat from them; but the hot air furnace is heated red hot or nearly so, and the current of air which passes over this superheated surface is burned, or loses a greater portion of its oxygen before it enters the room mingled with the heat, and it also has a chance to carry coal gas with it. The hot air of the house furnace passes directly from the furnace into the room.

Not so with the hot air heater of the incubator, for it will probably be heated to about 150° and the air which passes up the lamp flues into the heater never comes in contact with the eggs, but keeps passing out the vents and valves of the heater itself.

The temperature of the water in the hot water tank would be about the same if said tank was the same distance above the eggs.

It does not require a profound intellect to comprehend this, and persons who are deceived in the matter owe it to their own carelessness.

If hot air is foul air, why do not some of those wise fellows try cold air in the egg chamber of their incubators?

The man who spoils his own affairs through ignorance has our sympathy, even if he does not deserve any; but the man who undertakes to instruct others in matters which he does not at all comprehend, is almost a rascal.

The writer of a book for poultry raisers, published within the last three or four years—a man who poses as an expert in artificial hatching, and charges for his advice, says, in explaining his preference for hot water incubators over hot air machines:

"The reason for this, to our mind, is that the hot water gives more of a moist air, etc."

We expect to hear of this man writing a book for farmers, advising the use of butterflies and grasshoppers—the butterflies to make butter and the grasshoppers to make grass.

The advantage of the hot water tank in the incubator is, that when the water in the tank is heated it takes a long time for it to cool off, and it is thus a protection to the man who is too lazy to fill the lamps, for his eggs are not apt to cool off enough to kill the germs in them. But if he is too lazy to fill the lamps, he is apt to be too *tired* to go turn them down or add some cold water during the hot part of the day, and when he goes to look at the machine he finds the temperature 5° or 10° too high. How about the advantage then? If this happens on an extra warm night, or while he is away from home, what becomes of the hatch? Which do you think most disastrous, underheat or overheat? We say overheat.

The advantages of the hot air incubator are, that the heat or hot air can be turned away from and out of the hot air reservoir easily, quickly and automatically, without disturbing the condition of the egg chamber. The entire contents of the reservoir can be ejected by change of current, and replaced with cool air in a few minutes; and by action of the automatic regulator the heat currents are again turned into the reservoir and the temperature raised therein to the desired degree. This is done by the action of a thermostat in the egg chamber, controlling to a certainty and quickly the temperature within the heat reservoir. This is an utter impossibility with the hot water tank, so that when it becomes overheated, as is often the case, you must either turn down the lamp and wait for the water in the tank to cool, draw off some of the hot water and pour in some cold, open a large ventilator from the egg chamber and let the heat and moisture escape together, open the door and risk chilling the eggs, take out the eggs and sprinkle them with tepid water, or let the eggs cook. In a properly constructed hot air incubator (not one with single wall or thin walls) the heat is easily confined and controlled, being automatically supplied or cut off by the regulator, which acts as does the safety valve of a steam boiler. It also consumes less oil than the hot water machine.

The disadvantages of the hot air incubator are that the lazy man who cannot look at it once in 24 hours may let the lamps burn out and the machine cool down. Then, if somebody should deliberately put the lamps out, it would cool off, in the course of time.

But how about if somebody deliberately (or otherwise) turned up the lamp of the hot water incubator, would it not go to the other extreme?

The sage who says water in the heater tank is like the hot blood in the hen, knocks the "moist heat" theory in the head, for a setting hen does not sweat. Neither does a very warm dog. The overheated dog pants, and drops of moisture fall from his tongue; an overheated hen also pants.

But there is a time (we almost forgot to mention it) when the water tank gives moisture to the eggs, and that is when it springs a leak, which, in the majority of hot water machines, and especially cheap ones made by contract, is as likely to occur in the middle of a hatch as at the beginning or end—or may happen as easily at midnight as at daybreak. If you are on the spot just at the time and have another (empty) incubator heated up and ready, you can transfer the eggs to it, and send for a plumber or tinsmith at leisure.

Another peculiarity of the water tank is that when the lamp does go out unexpectedly, either by reason of neglect to fill or trim, or from the clogging and sticking of the lamp trips used on most hot water machines, and the water gets cooled off, it takes a long time to get it heated again; whereas, the temperature in the hot air reservoir can be raised from minimum to maximum in one-fourth the time.

From the opening of the incubator exhibit at the World's Fair, Chicago, until the day before it closed, we had three hot air incubators there in constant operation, hatching chickens.

In the hottest weather all the other incubators in operation in the building were obliged to put out their lights at some period of the day, but the lights on our machines were never extinguished, from start to finish of the exhibition, except when the gas fixtures in the building were being repaired or altered by the authorities. A new guard once extinguished the light on a hot water incubator, about six o'clock in the morning, but as the operator arrived soon after, no harm was done. This incident was made the stock of many good-natured jokes on the guard, and the nucleus of several funny newspaper paragraphs. A certain hot water incubator concern (not the one whose light was put out) has made it the subject of a grand fairy tale, in which they make it appear that *all* the incubator lights were put out, that the men who were operating the three hot air machines admitted that their eggs were ruined, but that no injury was done the eggs in their own machine. As the lights on our machines were never put out by a guard, we never had any complaint or admission of the kind to make, nor did we ever hear of any other lights than the one noted being put out by the guard.

MARKING EGGS.

The man who is in the poultry raising business for profit, will use every known means to make it a success. By a few marks upon the eggs he makes it possible to know all about each hatch, the causes of success or failure and the direct means by which to improve.

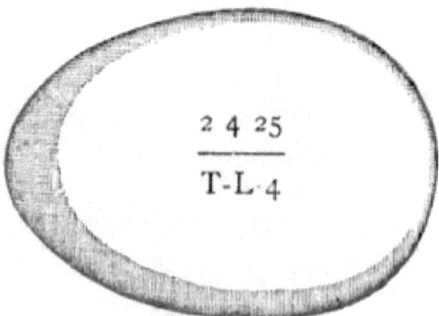

FIG. 14.

Fig. 14 shows one side of an egg marked for record. 2 indicates the month (February being the second) in which the egg was set; 4 the day of the month ; 25 the day of the month when it is due to hatch ; T indicates that the egg was tested on the 5th or 6th day of incubation, and that it contains a strong live germ, as shown in Fig. 1, in article on testing eggs ; L indicates that the egg was tested again on the 10th day ; and 4 that the air space is about the same as shown in Fig. 4, in article on testing eggs. A different figure would indicate that the air space on 10th day corresponded with that shown by a Fig. of corresponding number.

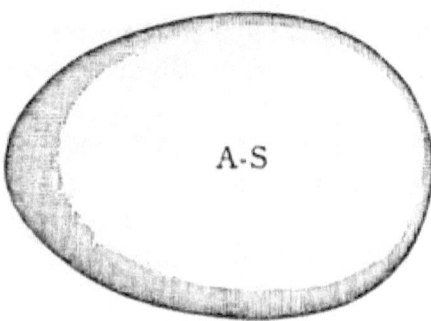

FIG. 15.

Fig. 15 shows the opposite side of Fig. 14. A should correspond with the name of the party or yard which furnished the eggs ; S indicates a thick shell. If the letter T were substituted it would indicate a thin shell.

We have said in another place that the different shells require different treatment to obtain the best results. You will notice that while, as a rule, dark shells are thicker than white ones, you will find some thin shells among the dark ones and some thick shells among the white ones.

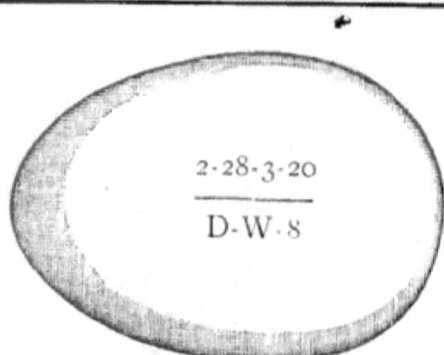

FIG. 16.

Fig. 16 shows a different egg marked on the same plan of Fig. 14. 2-28 indicates that the egg was set on 28th day of 2d month (February 28th); 3-20 that it is due to hatch on 20th day of 3d month (March 20th); D a doubtful or weakly fertilized egg when first tested on 5th or 6th day; W that it has been tested on 10th day; 8 that the air space on 10th day corresponds with that shown in Fig 8.

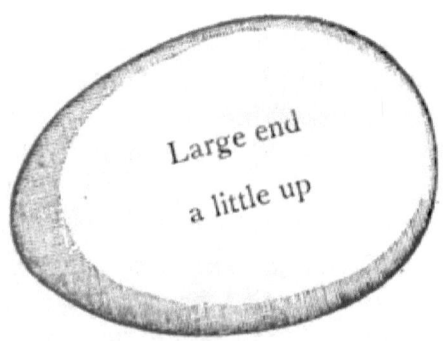

FIG. 17.

In placing eggs in the trays elevate the large end as shown in Fig. 17, so that the head of the chick will form in that end. If the head forms in the small end there are nine chances to one that it will never get out of the shell.

TABLE FOR KEEPING RECORDS.

The accompanying table will be found very useful for recording hatches, and such records will enable the poultryman to discover and explain intelligently the causes of his successes or failures in hatches; to anticipate and avoid the poor material, and to classify and properly treat that which he does use. In marking or recording air spaces make it as near as you can, but do not attempt to measure them all.

RECORD OF HATCHES.

Eggs from,	A	B	C	D	Day	Hygrometer, Wet Bulb	Dry	Spiral
Number of eggs,	100	100	100	100	1			
Variety,	L	M	B	PR	2			
Dark eggs,	50	—	100	100	3			
Light eggs,	50	100	—	—	4			
Thick shell,	60	5	95	96	5			
Thin shell,	40	95	5	4	6			
Fertile,	85	90	92	90	7			
Doubtful,	5	1	2	5	8			
Tested out,	10	7	8	5	9			
Age of egg,	7	5	6	5	10			
Air space 10th L.,	8	8	—	—	11			
Air space 10th D,	7	—	7	8	12			
Air space 16th L.,	4	4	—	—	13			
Air space 16th D.,	11	—	11	4	14			
Hatched strong,	84	87	80	90	15			
Hatched weak,	1	2	10	—	16		"	
Died about days,	1/10	2/10	5/10	2/10	17		"	
Died about days,	1/18	2/19	10/19	2/19	18			
Unfertile,	13	—	10	—	19		"	"
Per ct hatched F. E.,	85	89	80	90	20			
					21			

32

COOLING THE EGGS.

Cooling the eggs, or airing them, as it is generally termed, is a very important part of incubation, and careful attention to it will be repaid by an increased percentage and stronger chicks.

"I do not need to cool the eggs," says someone, "my incubator has all the ventilation they need." It may have plenty or too much ventilation, yet for best results the eggs should be cooled once a day, beginning on the second day and continuing to the eighteenth, inclusive.

The hen leaves her nest once a day, if allowed, and in exceptional cases where she does not do so voluntarily, she should be taken off once a day. The hen that leaves and returns regularly to her nest, hatches much better than the one that does not. In moderate weather in the spring the hen does her best hatching. She leaves her nest for a limited time and returns; the eggs do not get chilled, but are properly cooled.

In hot weather the hen is often driven from the nest by lice or mites. The eggs get plenty cooling, but do not hatch well. This is partly due to neglect of the hen and to a lack of vitality in the eggs. It cannot be all laid to too much cooling, because eggs will stand considerable exposure in hot weather. And it is so with eggs in the incubator. They may be left out much longer in hot weather than in the spring or winter. In early spring and winter the hen sits closer; she moves the eggs from centre to outside, and they are cooled quicker than in warm weather. When she leaves her nest once a day for food she returns quickly. The same course must be pursued with the incubator, *i. e.*, the eggs must not be exposed as long in cold as in warm weather.

Once a day, beginning with the second and ending with the eighteenth day, the eggs should be cooled to about 80° Fahrenheit, not cooler. This can be done after turning them in the morning. One soon learns to tell the degree of heat by laythe hand on the eggs or by holding an egg against the face. When the surface of the egg indicates 80° the inside is of course warmer.

The incubator should always be closed while the eggs are out cooling, for it is not desirable to cool the machine. When the hen leaves her nest she does not dive into the water or sit upon a cake of ice. When the eggs are out of the incubator it takes more heat to keep the egg chamber at the proper temperature, and the regulator, if it is a good one (and an incubator without a regulator is behind the times), will turn on extra heat, and when the cooled eggs are replaced, will turn on still more, automatically, which is turned off again in the same way when the egg chamber recovers its proper temperature.

Nine-tenths of the successful users of incubators cool the eggs; so do the manufacturers of incubators when they want to make a good hatch. Cooling the eggs is one of the important items in incubation, but not the only one, you will not succeed if you neglect the others.

TESTING EGGS.

This is a very important part of the business, and if properly attended to will throw a flood of light upon many perplexing problems in natural as well as artificial incubation. It not only elucidates but *proves* the truth or fallaciousness of our theories in the line of hatching.

Men are frequently heard to say that they never bother with testing eggs. That they cannot replace the unfertile eggs with others, and therefore nothing is gained. They are told by the best authorities that boiled eggs are not good food for chicks, and as for themselves, of course they would eat only fresh eggs. Then there is a risk of taking out hatchable eggs; so they run all the eggs through together. They say that they can break the unhatched eggs when the hatch is over, and see which were unfertile—and who cares whether they were or were not fertile if they did not hatch?

To those men we can only repeat. "Where ignorance is bliss 'tis folly to be wise."

To attain the best results it is absolutely necessary to test the eggs in process of incubation. If the eggs all come from one farm or yard, and they prove a large per cent. unfertile, weakly fertilized, or stale, you will notify the party from whom you got them, and he can look into the matter and rectify it, if he will, and afterwards serve you with vigorous fresh ones. If he will not do so, then you can avoid him, and procure better (or worse) ones.

If the eggs are from your own stock, and you know that they are fresh, and they prove unfertile or lack strength, you will know it, and can proceed at once to remove the cause, and thus save time, eggs, and complaints from your customers to whom you sell eggs for hatching.

If you have several yards, you should mark the eggs from each yard so that you can tell which are the best and which the poorest, and then treat the stock in each yard according to the requirements indicated by the testing of their eggs. There is a cause for each imperfection, and you should discover and remove it.

You may test your eggs this month and find them all right; next month they may be all wrong. suppose that you wish to set two hundred eggs, and get several lots of eggs from different yards or persons, to make up the number. One or two lots may be first-class, while of other lots nine-tenths are unfertile and the balance too weak to hatch. If the separate lots were not marked you would condemn the whole lot and the parties from whom you bought them; and if you did not test them, you would probably condemn the incubator or the hens.

In selecting and marking eggs it is well to avoid extremely large or very small ones, odd shaped ones and those with cracked shells.

In testing you can very often trace a number of unfertile eggs to a particular hen by a peculiarity in shape and a uniformity of size—that is where a considerable number of eggs of a uniform size all possess the same peculiarity of shape, you can be reasonably sure that they were all laid by the same hen. You can use that hen's eggs for market instead of putting them in the incubator next time (unless you remedy the defect in the bird), and leave room for better ones.

Among the causes of unfertile and weakly fertilized eggs are an insufficient number of cocks for the hens, or, which is just as bad, too many cocks to a yard or colony: old or worn out cocks, ill conditioned or debilitated cocks; overfat or aged hens; too close confinement of breeding stock, lack of green food, too much meat, forced egg production by the use of condiments; low vitality of stock, from neglect to feed properly or protect from the weather, or diseases.

Stale eggs are almost as bad as unfertile ones. After an egg is eight days old it begins to weaken, both the germ and the sac or tissues which envelope the yelk. The older the eggs are, the fewer the chicks that hatch, and the weaker are those which do hatch. The percentage of deformed chicks increases with the age of the eggs.

As the yelk forms no part of the chick, but is absorbed or taken into the chick just before hatching, and is its natural nourishment for the first twenty-four hours after hatching, it is important that the egg should be as fresh as possible when placed in the incubator. If the yelk should be stuck fast to the skin of the egg, the chick must die, although it may break the shell.

Persons have written to us, saying that they

have had from one to six chicks hatch on the second and third days after placing the eggs in the incubator; that they *knew* the eggs were perfectly fresh, having taken them out of the nests each day, and that they would like us to explain the cause of the "premature" hatches.

They were simply mistaken. There was no doubt that the eggs hatched at the times stated, but that they were all fresh laid could not be true, unless a miracle had been wrought. Human ingenuity has dispensed with the hen as an incubator, but it is and ever will be beyond human art or science to shorten the period of incubation. Newly laid eggs of certain breeds of vigorous fowls hatch from twelve to forty-eight hours earlier than eggs from some other breeds, or older eggs from the same fowls; but that is natural, and cannot be changed by man.

Those "premature" eggs had certainly been under a hen or hens, or subjected to a heat of at least 101° for from sixteen to seventeen days previous to being placed in an incubator.

Now, if a few of the eggs were sixteen or seventeen days old, we may reasonably presume that some of the others were nearly as old, and if those which hatched on the second and third days had live chicks in them, might not some of the others have had dead chicks in them, chicks that had started and, after being taken from the nest, died before they were placed in the incubator?

If these eggs were tested on the fifth or sixth day, any large chicks would show, and they would ordinarily be taken for bad eggs, if dead ; if alive, they would be taken for eggs previously started. But if the germs had died at any time between the thirty-sixth hour and the tenth day, an inexperienced person would probably call them fertile eggs and let them go, then wonder why they did not hatch. This happens more frequently than is generally believed. Such eggs are easily avoided by using the tester before setting the eggs.

Chilled, limed, scalded and cold storage eggs sometimes find their way into the incubator ; but persons should not allow themselves to be fooled so badly.

While the majority of persons who have good incubators make good hatches, there are some who would make decidedly better ones if they would just post up a little on a few important points which are easily learned by practice of simple and inexpensive experiments.

Few persons understand testing eggs properly. Some have a very imperfect tester ; some are unable to detect the fertile eggs closely—they cannot distinguish a dead germ from a live one, nor a weak from a strong one.

All eggs should be tested on the fifth or sixth day ; at this test all clear or unfertile eggs should be removed.

To become expert in testing eggs during incubation, it is necessary to have a good tester.

By the use of a good egg-tester and the engravings shown here, any person can, with a little practice learn to test eggs rapidly and accurately ;

the engravings show exactly how the eggs look in the tester.

To become an adept at testing eggs for hatching one has only to use a good tester, his eyes and a little judgment. Break in separate saucers (carefully) one which you suppose to be a good, strong, fertile egg; one which seems to be fertile, but

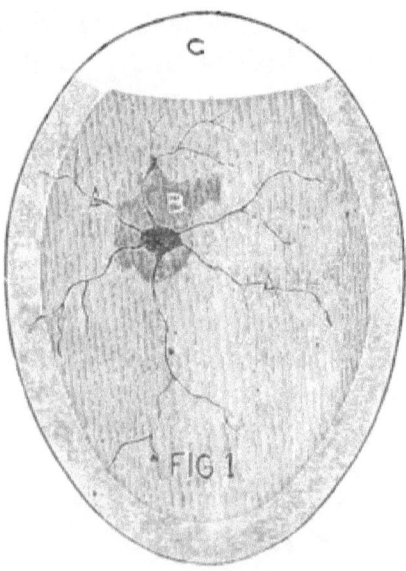

A STRONG FERTILE EGG,
On the fifth or sixth day, as shown with a good tester.

weak; one that is doubtful—that is, one which you cannot decide whether it is fertile or unfertile, and one that seems decidedly unfertile. Break one at a time, and examine it carefully, making note of it. This should be done on the fifth day, or at the first test.

A strong, fertile egg will, on the fifth day, (temperature having been kept at 102°, 103° or 104°) show a dark spot which will float and show veins running from it, looking somewhat like a spider; a weaker one will show a spot but is cloudy looking and muddled. The above are supposed to be fertile. Those which look clear are *unfertile*. Do not mistake the yelk for the germ or chick. All unfertile eggs are not perfectly clear. By breaking a few tested eggs and studying their contents, carrying in your mind's eye (so to speak) the appearance presented through the shell prior to the breaking; having broken an egg, say of the strong fertile ones, select another from the unbroken eggs, and see how it compares with the former. Then having opened a fertile but weak egg select another from the unbroken ones and see how well you can match the germ before you. Then break a few apparently clear and unfertile ones, and you will be surprised to find some fertile eggs among them if your tester is inferior, or you are careless. You will also be surprised to find how easy it is to train the eye to detect and classify minute things by a little systematic practice.

There is decided economy in this egg-breaking business, for it will save eggs and chicks in the end.

Do not blame the sitting hen or the incubator, unless you *know* that your eggs are *fresh as well as fertile*. We would not have eggs for hatching that are over eight days old at any price. We would not use them if given to us. We prefer them not over five days old, and would like them still better at or under two days old.

It is not hard to remember that *fresh eggs from healthy hens, fertilized by vigorous cocks*, MUST *be used* if we are to hatch a large percentage of strong, healthy chickens.

Fig. 1 shows a strong fertile egg as seen in the tester on the fifth or sixth day. B, the dark spot, is the live germ ; AA, are the blood vessels extend-

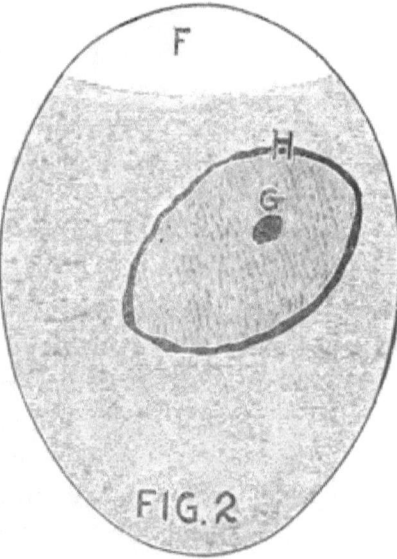

AN EGG TO BE DISCARDED

On the fifth or sixth day. Weak or imperfectly fertilized, as shown on the fifth or sixth day.

ing out from it. This germ B, is seen by placing the egg against the aperture of the tester and revolving it between the thumb and finger until the side on which the germ has formed comes nearest the eye. The spot B, will be seen plainly, often surrounded by a small cloud, as shown ; the germ at this time is quite lively, and can be seen to move up and down. This is a strong, fertile egg, and should hatch under a good hen or in a good incubator. In a well fertilized egg the blood vessels should show plainly, but the germ is not always seen as plainly, varying with the color and thickness of the shell and the power of the tester used. C, shows about the average air bulb in an egg on

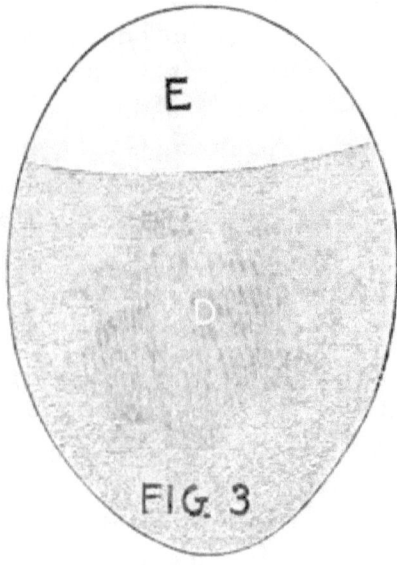

A STALE EGG.

As shown on fifth or sixth day: clouded, doubtful; many such should be broken.

the fifth or sixth day of incubation, though it may vary according to the freshness of the egg, and some eggs have larger air bulbs than others.

Fig. 2, shows a weak or imperfectly fertilized egg as seen in the tester on the fifth or sixth day. H, is an oblong or circular blood vessel which has started, but nothing more, there is no heart, nor any part of a chick started. This egg will not hatch, but will decay if left in the hatcher. G, shows a small dark spot, a weak germ, without blood vessels, only partially fertilized; it has died, after a start, and, of course, will not hatch. Both H and G, may sometimes be seen in the same egg. It will not hatch. F, the air bulb, may be seen in the same egg. The egg may be comparatively fresh, and yet show both G and H. See the following notes which explain why such eggs are found.

Fig. 3, shows a stale egg, a clouded egg, a doubtful egg. A stale egg is generally distinguished by the air space E, being very large on the fifth or sixth day, as shown in Fig. 3, though all stale eggs do not show a very large air space; but when an egg does show it, it is very good proof the egg is stale. When an egg shows a clouded, muddled appearance as indicated by D (which generally moves about when the egg is turned before the tester), it is certainly stale, and will not hatch. Do not confound the fresh egg which is not fertile with the stale egg; in an unfertile fresh egg you can see the yelk, which will look somewhat darker than the rest of the egg, but does not look muddled.

Fig. 4, shows a live egg on the sixteenth day. K, is the space occupied by the chick; the lines I and J, show the air bulb, which may be on top or at the side, as indicated by the respective lines. This is about the average air space on the sixteenth day, but it will vary according to the thickness of the shell and age of the egg when set; then some eggs are not as full as others. At this stage of incubation (sixteenth day) a live chick darkens the egg, except the air bulb, when seen with the tester, and by watching the line I or J, the chick may often be seen to move.

Eggs should be tested in a warm room, one tray at a time.

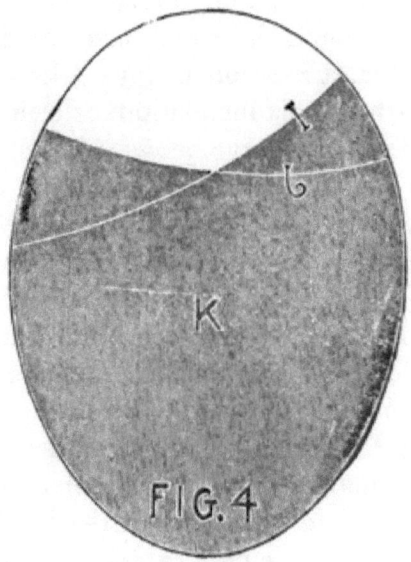

A LIVE EGG.

The air space on the sixteenth day.

The chick is harder to see after the seventh day, because the egg becomes more clouded by the growing chick.

NOTE. In regard to G, in Fig. 2, "a partially fertilized germ" means one that from one of several causes was not strong enough to live and grow. Among those causes are cocks that are too old, an insufficient proportion of male birds for females; old or debilitated hens, over-fat hens, too close confinement of breeding stock, etc.

Again you may find G (Fig. 2), among eggs which you believe or know are not over a week old, and ordinarily the eggs were good and fertile. It frequently happens that an egg will remain in the nest, while several, or maybe a dozen hens lay there, and the succession of layers keep the egg warm enough to start incubation, or it may happen that some eggs may have been subjected to a heat of 100°, in some warm place, unknown to or unnoticed by you. In either case, these eggs are taken from the nest or warm corner to a cooler place, and kept a few days, or over night, until a sufficient number has been accumulated to set, they become cold, and the germ dies before they are put under the hen or in an incubator.

In testing the first time, on the fifth or sixth day, a dead germ may be mistaken for a live weak germ, and if left in the incubator for three weeks would decay; so it is always best to test the eggs again on the tenth day, and remove all that have been marked doubtful and prove not good.

Some persons think it is just as well to leave all of them in until hatching is finished, but this is not right, the decaying eggs generate objectionable gases, and if broken are very offensive. A dead egg or an unfertile egg, does not contain the animal heat that live ones do, and are apt to have an undesirable effect upon the egg next to it, either under the hen or in the incubator.

An unfertile egg—one which has not been impregnated, and in which life will never start or develop—is clear when shown at the tester. This egg under the powerful lens of a first-class tester, will show the yelk, which must not be mistaken for a doubtful or fertile egg.

Use only the very best egg-tester.

HOW THE CHICKS DEVELOP.

Fig. 7 shows the heart and minute arteries and veins in a circle on the yelk which is enclosed in a thin sac. They are plainly seen by the naked eye when a strong fertile egg is carefully broken into a

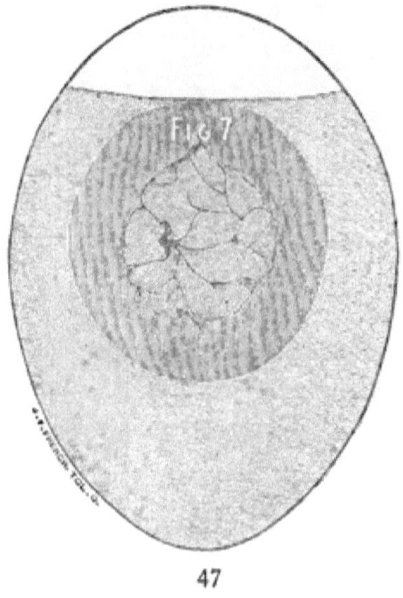

saucer or plate, after thirty-six hours' incubation. It should be done in a warm room, and in a strong light, when the pulsations of the heart will continue from five to ten minutes, and may be counted. Blood can be seen in the veins, but very faintly. The veins gradually surround the yelk. The chick derives nourishment from the yelk during incuba-

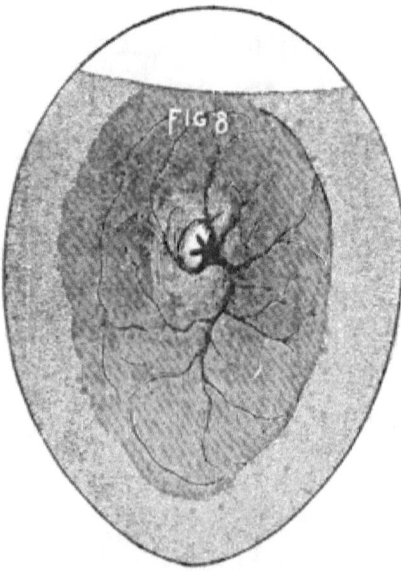

tion, and what is left of it is drawn into the abdomen just before hatching.

Fig. 8 represents the interior of the incubating egg on the fifth or sixth day, when the live germ can be seen with a tester moving up and down and around, and will float to the top when the egg is laid on its side. In testing, the large end of the egg is held up, as in Fig. 1, which shows exactly how the egg looks in the tester, through the shell.

Fig. 8 is seen with shell partly removed, or with the egg broken into a saucer.

Fig. 9 shows appearance on seventh to eighth day.

Fig. 10 represents the tenth day, when eggs should be tested the second time.

Fig. 11 shows the development on the fourteenth

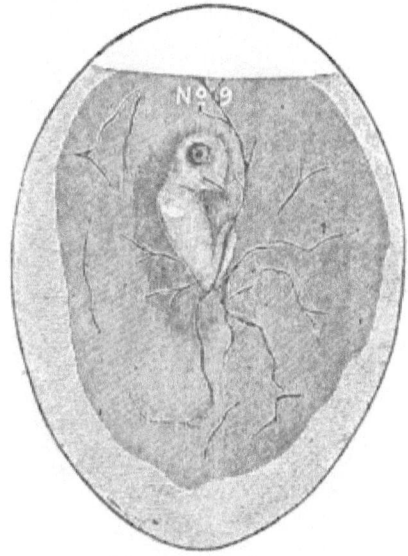

day. Notice the increased air space at the different stages.

Fig. 12, the sixteenth day.

Fig. 13, the eighteenth day, the yelk being nearly absorbed.

Fig. 5 shows the egg from the nineteenth to the twentieth day, when the chick is breaking the shell. At this stage the yelk should be entirely absorbed. The chick turns around in the shell, breaking as it goes.

Fig. 6 shows the shell parted and the chick ready to come forth.

As the yelk is the principal nourishment of the chick during incubation, it is desirable that the egg be perfectly fresh as well as well fertilized. The last part of the yelk absorbed is food for the chick for from twenty-four to thirty-six hours after hatching.

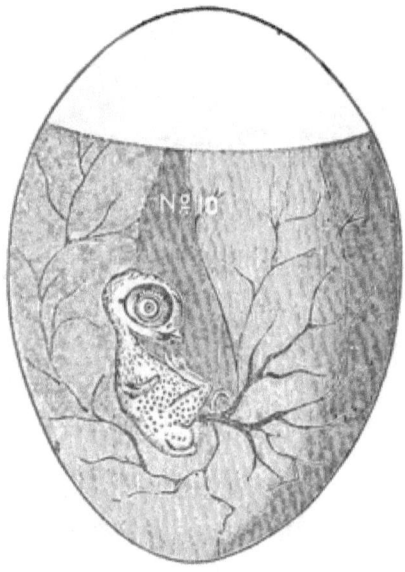

Stale eggs, though fertile, will not make hardy chicks; if they do hatch, the percentage will be small.

Break a few eggs that are not fresh, on a plate, and you will notice that in most of them the sac which confines the yelk will break and allow the yelk to mix with the white. A few which, being very carefully broken, retain the sac unbroken, present a mottled appearance and spread out flat, unlike the yelk of a fresh egg, which stands up and looks firm.

If the yelk is not in first-class condition it will not make a first-class chick. When eggs are stale many chicks will die in the shell on the seventeenth and eighteenth days of incubation, even when strongly fertilized.

FIG II

Deformed chicks are due to stale eggs, eggs from ill-conditioned stock, and overheat. Overheat will sometimes cause the chicks to break the shell before the yelk is entirely absorbed, and if you help them out in that case they will die.

Insufficient air space will prevent a chick from turning in the shell and from getting out.

When eggs have been crowded and some of the

shells are broken almost around yet the chicks do not break out with the majority in due time, you may then pull the shell gently apart, but leave the chick to free itself; for a chick which cannot free itself is not worth keeping.

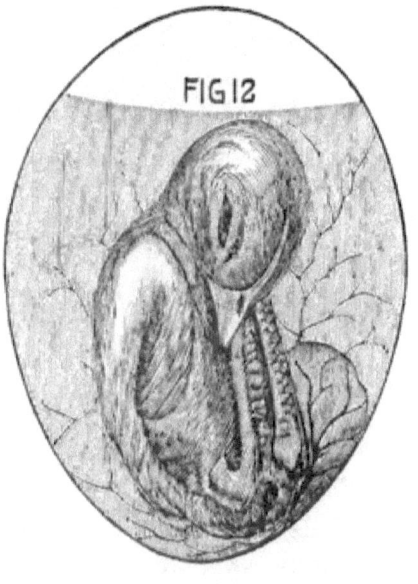

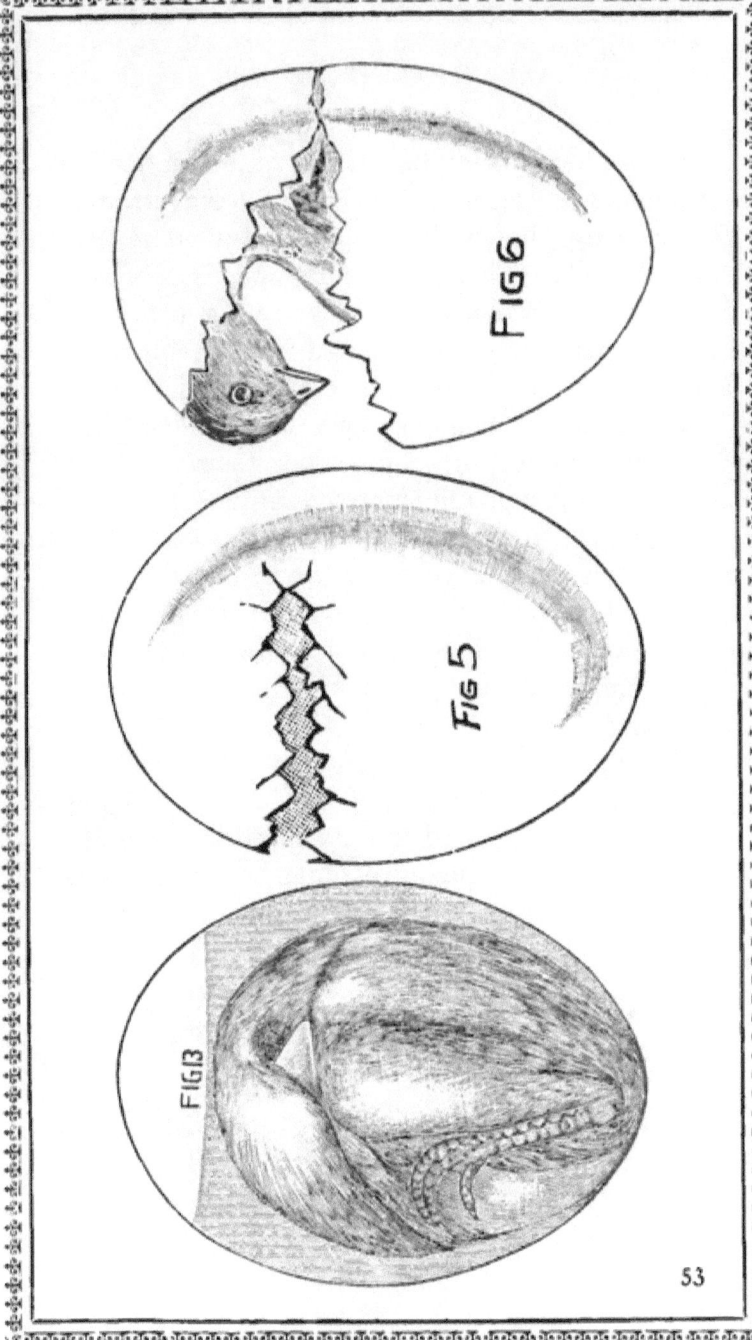

ANIMAL HEAT.

Animal heat during incubation is not noticeable until about the tenth day, though a fresh, strongly fertilized egg, having been subjected to a temperature of 103° for thirty-six hours, and then broken in a saucer, will reveal a live, pulsating heart, the beats of which may be counted, and it will make about sixty before it stops or dies.

As the animal heat increases less artificial heat is required to keep up the proper temperature in the egg chamber. On this account it is generally necessary to occasionally adjust the regulator a little the last week of incubation; but the lamp flame or flames should be gradually (a little at a time) lowered if it appears that you are using more flame than is necessary to supply just a little surplus heat.

Although it does not require as much *artificial* heat during the latter part of the hatch to keep up the right temperature, it should be distinctly understood, that the temperature must be kept up just the same as at the first part until every hatchable egg is hatched. [See When Hatching.]

WHEN HATCHING.

When the chicks are breaking and coming out of the shells, remember that the doors of the incubator should not be opened but twice in a day, to take out chicks that have hatched and are dry.

Also remember that when you take out a lot of chicks you take with them a lot of animal heat, and you should raise the lamp flame a little, for the temperature must be kept up to the same point until the hatch is finished, if you want the best results.

DEAD IN THE SHELL.

Why do chicks die in the shell; what is the cause of it?

This question is asked again and again in all the poultry papers. It is asked not only in regard to those that die in the mechanical or artificial incubator, but those also that die in the shell while under the sitting hens, ducks, turkeys, geese, etc. But the person who has just commenced running an artificial incubator, loses sight of this fact, and thinks that if some eggs hatch, every in egg the machine should hatch, or that certainly all eggs which start to incubate should bring out chicks.

While in a great many instances the majority of cases of "dead in the shell" may be justly charged to the incubator or the sitting hen, it is not always so.

Again, though there are first-class incubators which do hatch well, it must not be taken for granted that all incubators are good. There are good and bad incubators as there are good and bad hens and good and bad eggs.

The fact that some hens steal their nest and bring out a chick from every egg, or do nearly that well, is no proof that it was on account of having had their own way. Other hens steal their nest and only hatch one or two chicks; sometimes they fail to hatch any. A hen that steals her nest generally sits on the eggs laid by herself. If her eggs are strongly fertilized, and she is a good sitter and has a good place to sit, she will bring off a good hatch. If the eggs are not well fertilized she does not make a good hatch, but brings out perhaps six, two, or no chicks. The unhatched eggs may prove all unfertile, or most of the chicks may be dead in the shell.

What is the cause?

On the first event no impregnation; in the second, imperfect or weak fertilization. A bad sitter or poor incubator might cause the same result with good eggs.

When a good, quiet hen sits steadily on fifteen fertile eggs and hatches seven of them, is it not reasonable to suppose that the other eight must have differed somehow, in quality, at the beginning, or they, too, would have hatched? All having been subjected to the same conditions and treatment, why did not all hatch, or else all fail to hatch—all being fertile or containing the germ of life?

The answers to this question are legion; but most writers agree that it was lack of vigor in the germ, traceable to the parent stock, or to a malcondition of the laying stock, which produced the eggs. Had all the eggs failed to hatch we might reasonably suspect that the sitting hen had neglected her nest; but as seven of them hatched, the sitting hen is clear of blame, for the seven chicks could not have been produced without the favorable conditions for incubation to which they were subjected together with those which failed to hatch. Is it not plain that something was wrong with the eggs which contained chicks, in all stages of development, but failed to hatch? If the incubator (hen) was wrong, none would have hatched; if all of the eggs were right, all would have hatched. Now, the causes of unfertile and imperfectly fertilized or weak eggs are numerous, but easily removed or guarded against, provided we know what they are.

Too close inbreeding will make weak offspring. Inbreeding is excellent to a limited degree, but must not be carried beyond a few generations, if stamina and vigor are to be retained.

Over-fat hens do not produce eggs that will hatch well; no matter how good the male may be, the germs do not seem to receive the proper nourishment to develop strength to break out of prison, even if they grow to full size.

Stale eggs, however vigorous they may have been, do not hatch well.

Eggs may be both fertile and fresh, yet lack the vigor required to develop a chick.

Hens over two years old take on fat too easily, besides losing qualities requisite to good breeders. This is the rule. Of course there are exceptions; but you had better go by the rule than by the exception.

Some cocks retain a fair amount of vigor and procreative power after the second year, but nine out of ten do not.

If you want eggs to hatch well and to get the maximum profit from your poultry business, kill all the males and females at two years of age. Don't keep a fowl simply because it is fine looking. You cannot afford to keep simply ornamental birds in your flocks.

Fowls in too close confinement lose their vigor, and that, together with the practice of keeping fowls that are too old, is what causes nine-tenths of the "dead in the shell" cases which owe their origin to the breeding stock. Some people think a yard ten by twelve feet is large enough for the accommodation of a dozen fowls. They must have a reasonable amount of exercise.

As there are two classes of poultry raisers, there are two ways to effect a remedy.

The man who must raise his poultry on a limited area of ground, should keep fewer fowls. Is it not better to keep one hundred fowls from which you can produce eggs that will hatch from seventy-five to ninety-five per cent. of the fertile ones (seventy-five per cent. of all being fertile), than to house, feed and care for two hundred fowls to produce eggs of which fifty per cent. are unfertile, and only from thirty to forty-five per cent. of the fertile eggs hatch? Wriggle around it as you please, you cannot disregard this advice and succeed.

Those who have large tracts of land, but, because of keeping several breeds or varieties of fowls, are obliged to keep them in yards, should either enlarge their yards beyond (apparently) all reason, or at least beyond any size you ever saw before, and allow plenty of range for exercise and cleanliness, or reduce the number of varieties, and give each yard of fowls an extra *grassy* yard to pasture in for two hours each day; or, better still, keep but one variety and make kindling wood of your fences. Colonize your flocks on the Stoddard "no fence" plan, and you will have eggs that, with proper assignment and division of males and females (fowls), will show up ninety per cent. of fertility, and, in good incubators, produce from eighty to ninety-eight per cent. of strong, healthy chicks.

How do we know?

We have done it. The proof of the pudding is in eating it.

Now let us look at a few other causes of chickens dying in the shell; for you know it is quite possible to kill a vigorous germ or even a full grown chick by improper treatment. A poorly contrived incubator or a bad hen can easily destroy the life in the shell at any stage of incubation; or a careless or headstrong operator of a good incubator can spoil the hatch by what may seem to him a very insignificant deviation from the instructions of the maker of the machine.

Too much or too little moisture, heat or ventilation may ruin a hatch. Lack of moisture at the time it is needed, or excess of moisture when none is needed will injure or destroy life in the hatcher.

If the machine is deficient in any of these particulars, do not use it, but get one that you can depend upon.

You will also remember that eggs of various breeds vary considerably in shell, some shells being thin and porous, some thick, yet porous, while others are thick and dense or hard, and still others are hard and thin.

The treatment of these various shells is of importance, but will be discussed under the head of "moisture."

Chilling the eggs, especially during the last part of the hatch or while chicks are breaking the shell, causes many to die in the shell. See article on testing eggs.

Right here a little plain talk may be of some value.

In almost every poultry paper you will find complaints and queries about chicks dying in the shell. The correspondents, as a rule, wish to know the cause and the remedy. If the editor knows the cause he does not hesitate to prescribe a remedy; but there is where the trouble comes in. If we call in a physician to treat a case of illness, we give him an exact history of the case and all the symptoms and particulars; otherwise he could not prescribe to our advantage.

Are all poultrymen careful to do likewise when they ask advice from a poultry editor or expert, or incubator manufacturer?

We say no! Many of them do not know, or fail to mention the fact that a visitor "monkeyed" with the regulator, or that on one certain day they forgot to fill the incubator lamps and the lights were out for fourteen hours ; or that they forgot about turning the eggs, or left them out to air while attending something else, and they got chilled ; or that they forgot to put in moisture at the proper time ; that one of the children slipped into the incubator room and turned the lamp up or down, or put it out; or that a neighbor, who was looking at the machine, forgot to close the doors of the incubator ; or that the attendant accidentally set a tray, stick or some other trifle on the top of the machine in such a manner as to cover or rest on the closed valve of the heat escape, and the temperature got up to 110° before it was discovered. It is folly to omit or conceal these facts when they are known to the party who asks advice, because it cheats himself out of the chance of obtaining the remedy he seeks.

As we have intimated above, many poultrymen fail to give exact details, either because they have failed to see or notice some of them, or because they think they are not important ; and while we cannot advise them with as good effect as if we knew the real state of affairs, they cannot be said to deliberately deceive.

But what about that class of individuals who are so foolish as to deliberately lie when presenting their case of failure to the maker of the incubator, which they happen to be using? What help can a man hope to get, who, having bought a lot of store eggs which produce forty per cent. of weak chicks and sixty per cent. dead in the shell, in his incubator, into which he placed *all* the eggs he bought, when he writes to the maker of the machine as follows: "My hatch was forty per cent. of sickly looking chicks, and the balance of the eggs had dead chicks in them—some seeming to have died on the tenth day and others at various stages of development, many being full grown and just ready to hatch. I cannot understand it, as the same eggs placed under hens hatched ninety-eight per cent. I followed your directions to the letter."

Now if the manufacturer *knows* from years of severe test that his machine will always hatch a good per cent. of *hatchable* eggs, *every time*, when operated by his directions, he also knows to a certainty that his correspondent either failed to operate the machine as directed, or that his statement about hatching some of the same lot of eggs under hens is false. He knows it as certainly as the painter would know that we were speaking falsely if we told him that by mixing equal proportions of red and blue we produced green. Suppose we did write such a statement to a painter, and asked his advice, assuring him that our neighbor had used the same colors and produced purple. Would he not know beyond a possibility of a doubt that either we were color blind, or liars, or fools, or that we thought him a fool when we presumed that he would not know any better than to believe the statement.

All the advice he could give us would be to discard the yellow (which we called red) and replace it with red if we wished to produce purple by mixing with blue. And so would the maker of the good incubator have to tell his correspondent to hunt up a lot of fresh eggs from a place where the breeding stock is healthy, vigorous and mated with good proportion of males.

Other parties who have really first-class hatchable eggs, imagine that they can improve on the directions which are sent with the incubator (not having given those directions one fair trial) and will run the machine to suit themselves, and when they kill a lot of chicks in the shell, condemn either the eggs or the incubator. If they write for advise, they tell the truth about the eggs, but omit to mention the fact that they paid no attention to directions. Some of them, we regret to say, will even assert that they did follow the directions.

These persons are as foolish as the man who, suffering with cholera morbus, would tell the physician that he had toothache.

PERIODS OF INCUBATION.

Chickens, twenty-one days ; ducks, twenty-eight days ; geese, thirty days ; turkeys, twenty-eight days ; Guinea fowls, twenty-five days ; pea fowls, twenty-eight days ; pheasants, twenty-five days ; partridges, twenty-four days ; ostriches, forty to forty-two days.

In connection with the above table we should remember that a strictly fresh laid egg will hatch several hours earlier than a stale one, and that the fresh eggs of some breeds of the same species of fowls will also hatch from twelve to forty-eight hours earlier than those of other breeds, under the same conditions.

Some hens sit closer on the eggs and keep the temperature more regular than others do. These hens generally bring off their chicks in due time. Other hens are poor sitters, do not settle down on the eggs nicely, and frequently vacate the nest. Such hens are sometimes one or two days later than schedule time in completing their hatch.

The same principle applies to incubators. Those which maintain an even condition of requisite heat, moisture and ventilation, will, if the eggs are all right, complete the hatch on time ; while the irregular machine, now hot and then cold, now dry and then moist, will of course be behind time.

MOISTURE IN HATCHING.

How much moisture should be used in an incubator?

Why not ask, " How much lumber will it take to build a house?" The question is as comprehensive.

The question of how much moisture should be used in an incubator, never has been fully and correctly answered.

"Oh, yes, it has!" exclaims somebody, "Mr. A. says he uses no moisture at all ; B., none until the fourth day ; C., none until the seventh ; D., none until the tenth ; E., until the sixteenth ; F., until the nineteenth. G. uses water surface equal to three-fourths the area of the egg chamber ; H., five-eighths ; I., one-half ; J., three-eighths; K., one-fourth; L., one-fifth ; M., one-twelfth. N. will evaporate twelve quarts of water in the egg chamber during one hatch ; O. will evaporate one pint.

Any one of the above may be exactly right for some incubator, at some particular time, in some certain place, climate or season, and with certain kind of eggs, and it may never, in the experience of the operator, be right again.

How then are we to know the amount of moisture to use? If it is right at one time, why not always?

It might be right always with one particular incubator and the same kind of eggs, if the temperature and humidity of the outside atmosphere were always the same; but you know that is not the case. You cannot find two periods of three weeks each in which the twenty-one days of one will even average the same as the twenty-one days of the other, in any location, in a lifetime. Therefore there must be a vast difference in hatching, both with incubators and with hens.

For instance, we take a hot water incubator which has an opening valve for the escape of hot air from the egg chamber when the heat rises above a given point. No matter what size the opening may be, how large or small the moisture pans, or when the moisture pans are filled, on a hot day or when the lamp flame has been a little too high, this valve or escape will open, and from ten to twenty times as much air will be circulated through the egg chamber as there was the day previous or on a cooler day, and the next day it may be more or it may be less.

If there is no water in the moisture pans on the very warm day, then there is from ten to twenty times the amount of evaporation from the eggs; if the pans are filled, you will not have the same amount of moisture when the valve is open as you have when it is closed.

Let us suppose, for argument sake, that we have more moisture in the egg chamber with the valve closed than with it open; then, when the valve opens to cool off the egg chamber, the moisture escapes with the heat, and we have from ten-fold to twenty-fold reduction of moisture on the hot day, or when the machine is overheated.

But suppose we say that when the valve is open there is from ten to twenty times as much water evaporated as there would be with it closed, and that this water or moisture passes over the eggs; then we have practically from ten to twenty times as much moisture with the valve open as we have with it closed.

Clearly we cannot have the same amount of moisture with an open valve as with it closed—no matter which condition of valve gives the most, and, as this valve may open and close once a day or fifty times a day, how is it possible to maintain an even condition or degree of moisture in an egg chamber which is thus "regulated?"

Then why use a moisture gauge?

A moisture gauge of the best make will show that the humidity of the egg chamber fluctuates as we have stated, under the said conditions.

Wherever there is a circulation of air in the egg chamber—and there *must be* a circulation of fresh air to hatch successfully—there will be *some* variation of humidity, because the humidity of the outside atmosphere changes, and it is this outside air which furnishes the fresh air for the egg chamber; but with the valve closed the variation is reduced to a minimum degree.

The modern hot air incubator has no use for the big trap-door valve opening from the egg chamber of most hot water machines, because the heat is controlled in the heater or hot air reservoir before it reaches the egg chamber; but it has a circulation of fresh air in the egg chamber by means of bottom ventilators and small outlet at top. These inlets and outlet are *always* open and *always the same*, so that the variation is reduced to a minimum, being about equal to the variation in the best hot-water machines with its valve closed.

The variation of humidity in the best hot-water machine with closed valve and in the best hot-air machine just as it always is, is from ten to twenty times less than when a large valve is being opened and closed from the egg chamber every now and then, as described, and the causes for the minimum variation which cannot be entirely overcome are two, the variation in humidity of outside atmosphere and the variation of temperature of outside air.

We all know that heat ascends. As long as the air in the egg chamber is warmer than that outside, the heat of the egg chamber will seek an outlet, and the colder the outside temperature the faster the current of air will circulate through the incubator (the greater the volume of air that will pass through in a given time), but the moment the outside temperature gets as warm as that of the egg chamber the current ceases. Then it follows that more air passes through the egg chamber on a cold day than on a warm one, consequently there is more moisture supplied from the outside on a cool, damp day than on a warm, dry one.

When you look at all these facts and think of the difference in incubators, the wonderful variety of climates and altitudes on this vast continent, the different kinds of shells which envelop the eggs—thick, hard shell, thin, hard shell, thick porous, and thin porous—each requiring different treatment (the hard shell requiring more moisture than the porous shell), you will appreciate the difficulty of giving a direct answer to the question of "How much moisture should be used in an incubator?" It is impossible, in the ordinary instructions which accompany an incubator, to give directions which will fit every case in every locality, therefore the incubator manufacturer who would conscientiously perform his duty to each of his patrons must have an *actual* knowledge, which can only be acquired by *actual experience*, of the action and requirements of his own incubator in the various altitudes and climates of this country under the various conditions of the seasons. Having gained this experience he can then give directions with each incubator to suit the locality to which it is sent. These directions would (or should) be such as will give the best average results with that particular machine in that locality.

For the benefit of new acquaintances who may wish to inquire if we practice what we preach, we will state that during the years of our experimenting we paid our compliments personally to the chief points of interest lying between the Atlantic and Pacific oceans, the great Northern lakes and the Gulf of Mexico—not making a flying trip, but spending months and sometimes years in climates and altitudes of peculiar interest to poultrymen who practice artificial incubation and brooding.

As we have said, follow the printed directions sent with the incubator, and then at your leisure study and experiment for yourself. If you find that the manufacturer's directions give satisfactory results, the knowledge you have acquired from study and experiment will enable you to see why they do; and if, on the contrary, they do not satisfy you, you may then be able to improve upon them.

Someone will say, "What a lot of fuss about moisture! Let me give you the whole thing in a nutshell. Find out just what degree of humidity is needed in the egg chamber for each week or day, make slide covers for your moisture pans, place a hygrometer or moisture gauge in the egg chamber and hang up your moisture schedule beside the machine. When you want more moisture slide open the covers, and when you want less, close them. Isn't that simple?"

Yes, dear friend, wiser heads than yours or ours thought of that years ago, but it would not work then, and it will not work now.

Why?

For various reasons; among them: the Great Ruler of the Universe will not permit us to slide the covers of His moisture pans; and while we are obliged to circulate fresh air in the egg chambers of our machines, we are obliged to have it more or less humid or dry, just as it comes from the breath of nature.

The hygrometer is useful to experiment with, provided it is a good one, but few of those which are sold to poultrymen are reliable.

Still someone says, "Well, I know that the humidity of the atmosphere varies some, but I still believe I can work it with the moisture gauge and the sliding covers on moisture pans."

Very well, we will ask you for one demonstration, and if you make that satisfactory, we will ask for one or two more—but one will probably be all you want at a time.

Let us suppose that you conclude that you want thirty degrees of moisture in the egg chamber the first week, thirty-five the second and part of the third, with ninety degrees from the pipping of the first egg? All right. We will take for granted that your gauge is correct. Well, here we are at the beginning of the first week. You have not yet put any water in your pans but your moisture gauge indicates sixty-five degrees of humidity, and your thermometer one hundred and three degrees of temperature. What is the matter; why don't you reduce the humidity? You place another moisture gauge in the room where you operate your incubator, and you find that the humidity there is ninety degrees. You hang a gauge in the open air out of doors and it registers ninety-five degrees. You only want thirty degrees in the egg chamber; how are you going to reduce it to thirty?

There are some places in which, at certain times, some kinds of eggs can be hatched without additional moisture, in certain incubators, but the attempt would result in failure at other seasons.

It is surprising how little some manufacturers of incubators know about moisture. When you attend a show where incubators are on exhibition question the several exhibitors on moisture. The machines are generally managed by the manufacturers or inventors—or the purchasers of some almost defunct patents.

While exhibiting at the World's Fair, Chicago, '93, we were astonished at the replies of some of the manufacturers and exhibitors when asked "why do you use moisture? What is it for?" One said, "Oh, the hen sweats and moistens the eggs, that is the reason." Another said, "We imitate the hen; she goes off in the grass and gets the dew on her feathers and dampens the eggs; we must supply it." Another said, "We use moisture to *rot* the shell of the eggs." Another, "The evaporation of the water purifies the air."

We supposed that everybody knew that moisture is used in an incubator to prevent undue evaporation of the egg, and to keep the skin which lines the shell from becoming dry and tough while the chicks are breaking the shells.

HATCHING DUCKS.

Duck eggs require about the same treatment, during incubation, as hen eggs, except that the addition of moisture is deferred one week. Ducklings are longer getting out of the shell after it is broken than chicks are—from twenty-four to forty-eight hours is the time they require to work their way out. If, after waiting forty-eight hours after the eggs are pipped and ducklings are not free, you may help them out gently. There is not as much danger in thus helping them as there is in assisting chicks out. A chick which cannot free itself from the shell is not worth saving, but a duckling is. Ducks and ducklings should have water when eating and water to drink at all times. Keep ducklings from bathing or getting wet until feathered. Celery fed to ducks one week before fattening is supposed to improve their flavor. In killing ducks hang them up by the legs and extend the head or bill to prevent soiling the feathers. With a sharp knife cut across the back part of the throat and up into the brain. Ducks are easier picked when scalded. After picking, put them in ice water until thoroughly cooled. N.B.—Make the drinking vessel deep enough for the duckling to wet his nostrils, or they will become clogged with dirt or soft food and make him sick.

HATCHING GEESE.

Geese develop as well in the incubator as under geese or hens, but goslings are not very dexterous in breaking the shell, hence many are lost, because few persons know how to help the process. When the hatch is due hold the egg in a strong light and try to see where the gosling is tapping the shell, which you can often do. Make a small hole with the point of a penknife, and if no blood oozes out make another hole in the large end of the egg, and chip away the shell between the two points first, and then gradually break away enough to free the bird. If you cannot see where the bill lies, put the egg in warm water, mark the spot which lies uppermost, and make the first incision there. With gentle care nearly all may be saved. Feed goslings the same as ducklings, adding green food early, and keep out of the water until feathered.

HATCHING TURKEYS.

There is no difficulty in hatching good turkey eggs in a good incubator. Treat the eggs precisely as you would hen eggs, except that the moisture must not be added until a week later than directed for hens' eggs. There is a general opinion that young turks are hard to raise, but the great difficulty is that few persons know how to treat them, and others do not have a large range for them from the time they "shoot the red" (the head begins to turn red) until they get their growth. Young turkeys must be kept dry until they show the red. Running in wet grass, or exposure to rain, will retard their growth and prove fatal to many. Brood them as you would chickens. Their food for the first week should be stale bread crumbs (not sour) soaked in milk. Broken water crackers soaked in milk for a variety. Give them all the sweet milk they will drink (by "sweet" we mean *new* milk). After the first week give them all the clabber they will eat, but do not scald it. In addition give once a day a small feed of well cooked corn meal, sometimes in the shape of mush, and again as baked corn bread. By draining the clabber through a cheesecloth strainer it becomes nice and crumbly, and is easily picked up. Keep them in brooding yards until about eight weeks old, then give as much range as possible. After eight weeks give no sloppy food, but good grain—corn or buckwheat at night, and a variety of food in the morning. They will gather at least half of their living in the insect season if they have good place to forage. A turkey is unhappy in close confinement and will not fatten in a pen like other fowls. Liberty and good feeding will give the weight to a well bred bird.

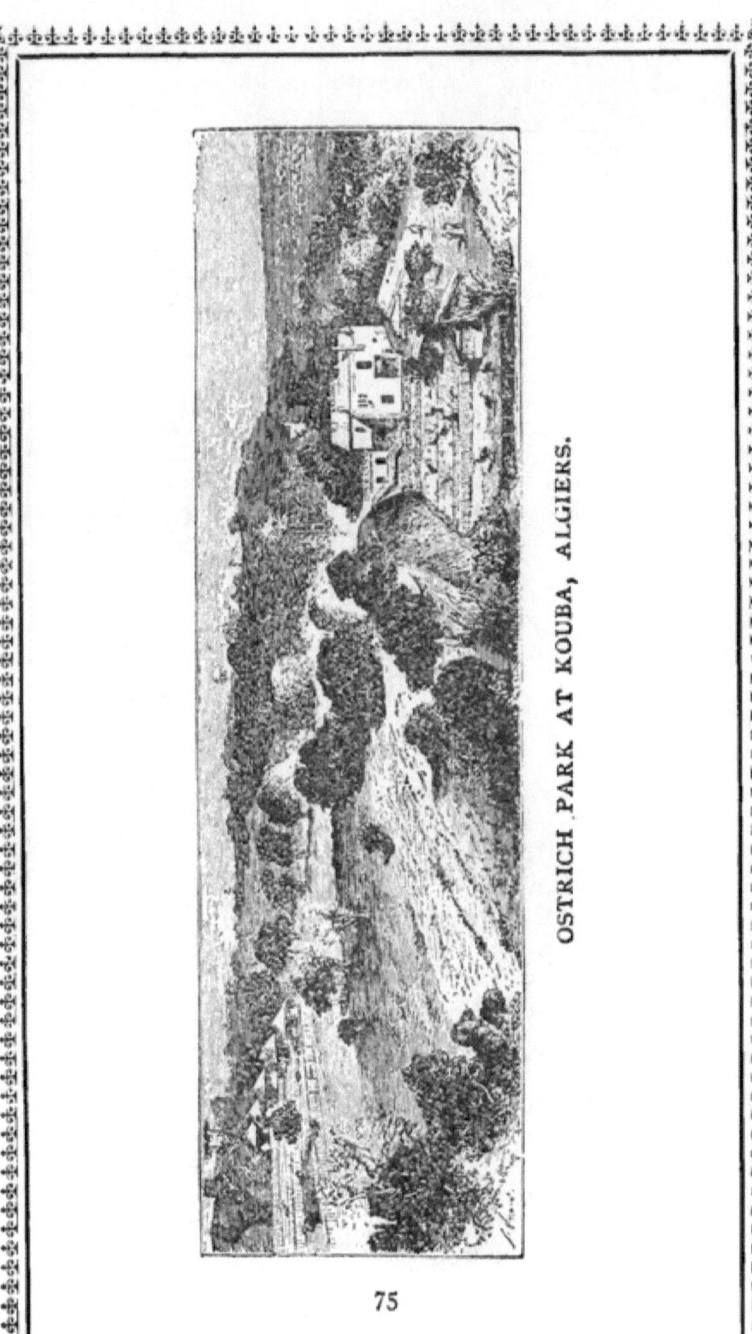

OSTRICH PARK AT KOUBA, ALGIERS.

PARK FOR YOUNG OSTRICHES.

PARK FOR SITTING OSTRICHES.

HATCHING OSTRICHES.

THE introduction of artificial hatching has added materially to the profits of the ostrich parks and farms of Africa, Asia and America. Sitting is injurious to the valuable plumage of the parent birds; and then the eggs may be used for hatching without consulting the convenience of the layers. The period of incubation is from 40 to 42 days, and little more care than is required in hatching chickens is necessary in hatching ostriches. The young birds are as tender as young turkeys, and should be kept in the brooding house until the sun has dried the grass. They must also be returned before the dew falls. They need shade in the heat of the day, but the more sunshine they get, that is not too hot, the better they will thrive. The eggs weigh from 3¼ to 3¾ pounds, and one before us measures 15⅜ inches by 17¼ inches around each way. They are palatable and wholesome when boiled, but are too precious for ordinary table use.

The African ostrich is superior in size, weight

and quality of plumage to the Cassoway of New Zealand, the Rhea of South America, or the Emu of Australia, and is the kind bred on the Southern California ostrich farms at Anaheim and Fall Brook.

The first successful ostrich farms were those of the Cape in Africa, which started about 30 years ago. Later Madam Carrière established a series of ostrich parks at Kouba, Algiers, views of which were drawn by M. Louis Say, and which, through the courtesy of Messrs. Munn & Co. of the *Scientific American*, we reproduce here. Some idea of the development of ostrich culture may be drawn from the fact that the number of adult birds on the Cape farms in 1865 was 85; in 1875, 32,000; in 1879, 160,000.

E. & C. VON CULIN,
Publishers of
"*The Art of Incubating and Brooding.*"
Price $1.00.

DELAWARE CITY, ———, 1894.

Dear Sir:—We wish to illustrate the different kinds of incubators, brooders and poultry appliances. Will you please send us cut or cuts of your machine or machines, together with directions for operating and your catalogue. No cut to be larger than 3 by 3 inches.

Please be prompt, as we are ready and waiting.

Yours respectfully,
E. & C. VON CULIN.

The above letter was mailed to forty manufacturers of incubators. We presume they all received it, as none were returned, though our printed address was on each envelope. We waited a month. Five responded; some of them sent cuts, but only *one* sent directions for operating. We give these facts as an answer to those who may wish to know why we have not described all incubators, as well as a few. Without directions for operating, the simple picture of the exterior of a machine, is no more than you can see in the advertisements in poultry papers and in the catalogues of the manufacturers.

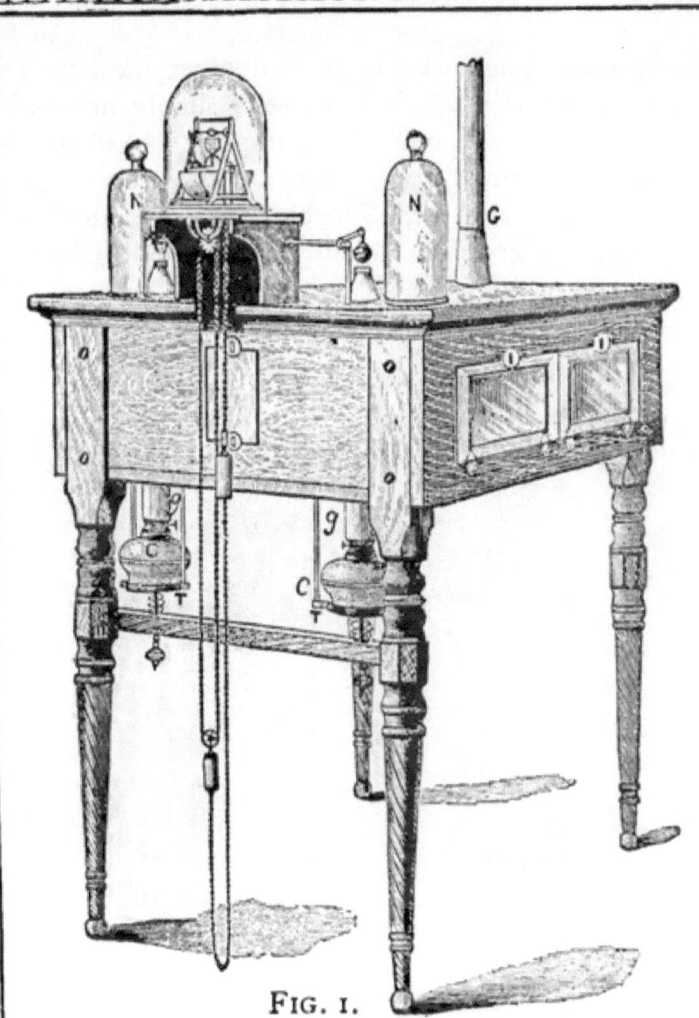

Fig. 1.

THE THERMOSTATIC INCUBATOR.

The Thermostatic Incubator was patented July 31, 1877, by E. S. Renwick. It was one of the finest hot water incubators made in its day. It is

manship. Then why is it that other incubators are offered at such low prices? Simply because the material in one of Mr. Campbell's incubators costs more than many of the gaudy rattle-trap machines are sold for. The craze to get something for nothing creates a lively market for worthless so-called incubators.

THE EUREKA INCUBATOR.
Manufactured By J. L. CAMPBELL, West Elizabeth, Pa.

The Improved Simplicity Hatcher consists of an egg chamber enclosed on four sides, top and bottom by double walls, six inches thick, four inches of which are packed with a light non-conductor of heat. Above this egg chamber is a heater (hot-air), also enclosed by the double walls, supplied by lamps through fire-proof conductors. In the centre of the egg chamber and on a level with the eggs is a thermostat which controls the temperature in the egg chamber and at the egg level. It is connected by a brass rod and lever with a valve on top of the hatcher, operating in such a manner that just as soon as the temperature in the egg chamber reaches 102° (or such degree as it may be set at) the valve begins to open, turning the

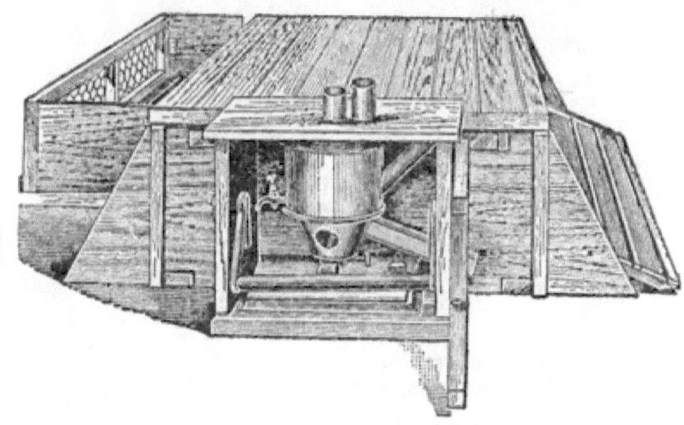

THE EUREKA BROODER.

heat currents away from and over the heater, instead of into it, and drawing cold air into the heater; and when the temperature in the egg chamber starts to fall below 102° it cuts off the cold air and turns the currents of hot air into the heat reservoir again, thus making a commutual or reciprocal action between the mechanism of the egg chamber and that of the heater, by means of which the temperature is kept absolutely under control. The ventilation is from the bottom, fresh air being diffused and escaping automatically through a minute tube in the top, without perceptible draught, and unaltered by the action of the regulator. Moisture is also supplied from below, at the time and in quantity suitable to the location.

Fig. 1. The general exterior of hatcher, showing

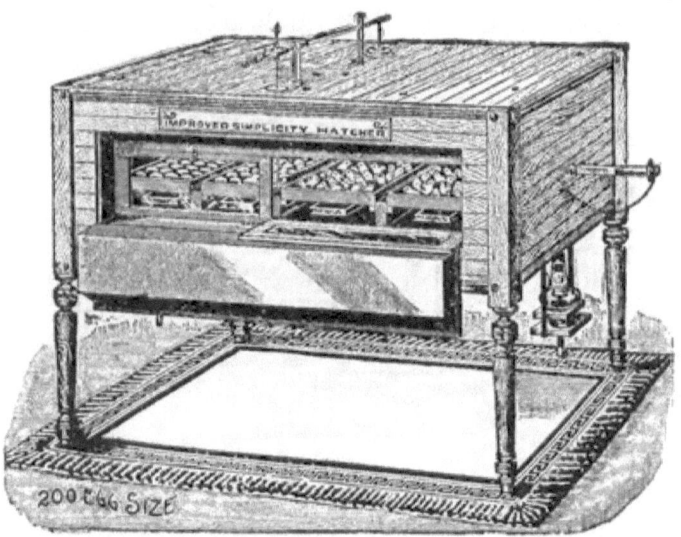

FIG. 1.
IMPROVED SIMPLICITY HATCHER.

the position of lamps, the glass door and the double packed outside door. The rod on the right shows the thermometer drawn out to observe the temperature; when replaced, the bulb is in the centre of the egg chamber and level with the eggs.

Fig. 2. Is a view in perspective, showing the body with top and outer walls removed.

Fig. 3. Is a longitudinal section on the line XX of Fig. 4, showing the thermostat and its connections.

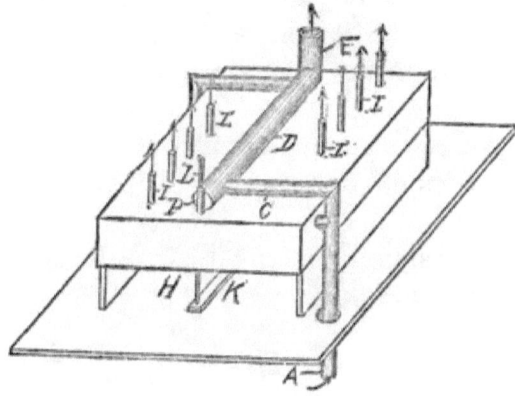

FIG. 2.

Fig. 4. Is a view in cross section on the line YY of Fig. 5.

Fig. 5. Is omitted, as the four figures shown are sufficient to explain fully.

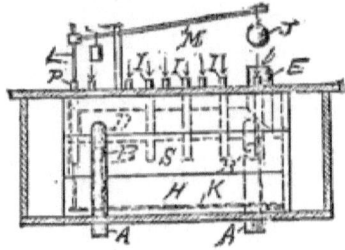

FIG. 3.

Similar letters refer to similar parts throughout the several views.

AA are inlets for conducting hot air into the heater S through pipes BB. When E is opened, it allows the hot air to pass up AA and CC, through D, and out at E, making a draft which draws hot air out of the heater S at BB, at the same time drawing cold air into the heater S at IIII.

D is a discharge pipe of double the capacity of C, and carries off hot air from CC out at E when E is open.

E is the main outlet for hot air, S is the hot air heater five inches deep.

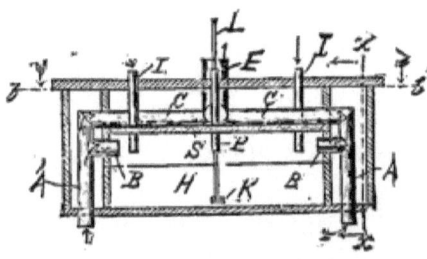

FIG.-4.

H is the egg chamber.

IIII are tubes running through the top of heater S, through which cold air is drawn into heater S, when E is open.

K is a thermostat in the egg chamber and is on a level with the eggs.

L is a metal rod connecting the thermostat with the lever M.

J is a ball or cover suspended at the outer end of the lever M, and is made to open or close the outlet E by action of the thermostat K.

P is a tube running from the egg chamber H through the heater S, and through which the rod L passes. A *constant* discharge of air flows through this tube, from the egg chamber. It is *never closed.*

X is one of the lamps, two being used at diagonally opposite corners of the incubator.

YYY is a five-inch space between the inner and outer wall of the improved incubator and is packed with mineral wool and granulated cork. All of the hot-air pipes are incased in asbestos or mineral wool.

DIRECTIONS FOR OPERATING
THE VON CULIN IMPROVED SIMPLICITY HATCHER.

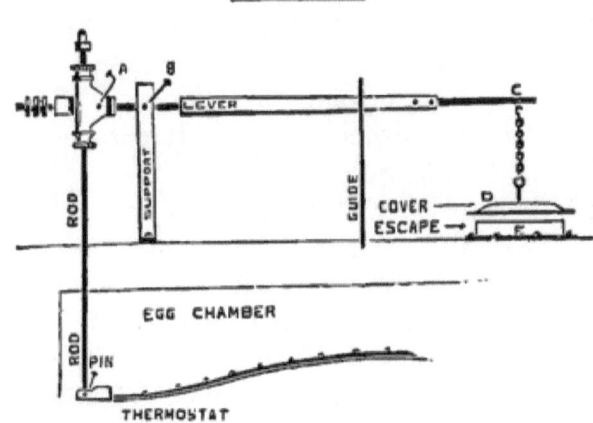

Light the lamp or lamps, turn the flame to ordinary height, with regulator as shown above, except that the perpendicular rod which connects the lever with the thermostat in the egg chamber by means of two brass pins must be left out (not connected) until the temperature in the egg chamber rises to 102°. Then insert the rod, and connect as shown in the illustration. When heating up the machine let the cover J rest on the escape E. To raise the temperature of the egg chamber, raise the two nuts on the perpendicular rod; to lower the temperature, lower the same nuts. Run the machine between 102 and 103°. The cover should be raised from the escape about one-sixteenth of an inch at 102°. This will give a surplus heat which will not rise above 103°, but if the temperature of the room should fall 40 degrees, and the temperature of the egg chamber should fall one-half a degree, the closing of the escape valve will change the current of the escaping heat at once, making it impossible to cool down; at the same time the cutting off of the excessive outward draft increases the flame of the lamp, without turning the wick or using any device upon the burner or wick. It is done by the change of draft alone.

The thermostat in the egg chamber opens and closes the escape valve to decrease or increase the heat in the heat reservoir, but does not open or close any valve in the egg chamber. The brass "guide" straddles the lever. The "support" holds the lever by a brass pin on which it works. The nuts on the perpendicular rod are to adjust the regulator. The nuts on the end of lever are to balance it to the makers' adjustment, and *must not be moved*. Run the machine between 102° and 103° —let 102° be your low point,—and 103° your high point. Run it empty for 24 hours to get well regulated, then put in the eggs. Always fill the lamps in the morning, if possible. After the eggs have been in 24 hours, turn them twice a day with the extra tray. Fill the moisture pans on the tenth day, unless otherwise directed for special location or altitude. Test eggs on fifth or sixth day (see "Testing Eggs"); test again on tenth day. If you wish to gain knowledge, test again on sixteenth day. Test one tray at a time. If the room is very cold take them into a warmer room to test them. Do not have a fire in the room where you keep the incubator, but have the room well ventilated at all times. Never turn up the lamp flame when the cover is raised from the escape. If the cover is raised high, say $3/8$ inch or more, and the temperature is right, you are wasting oil, and should lower the flame. If the cover is down and the temperature is too low, raise the flame. Always close the doors of machine when you take eggs or chicks out. You will not need to look at the hatcher more than twice a day, night and morning. After the first day cool down the eggs to about 80° or 85° once a day (when turning), until the chicks begin to break the shell, then do not turn or cool them any more, but place the "chicken guards" on the trays. Take out the chicks morning and night (only those that are strong enough) and place in a brooder. Do not open the machine often when the chicks are hatching. Remember that when you take out a lot of live chicks you also take out animal heat, and turn the flame up a little more. After filling the water pans see that they do not get dry. Trim the lamps once a day and keep the burners clean.

THE SIMPLICITY COMPARTMENT HATCHER.

This machine is like the Improved Simplicity in every respect except the position of the lamps, and is divided into from two to ten compartments, each compartment having a separate heater and regulator, and may be run independent of the other compartments. One, two or all the compartments may be used at a time. Persons who cannot furnish a large number of fresh eggs at one time can fill one or more compartments and start them to hatching, and fill up each of the remaining compartments at their convenience, thus setting *fresh eggs* every time. They may also be used to advantage by persons who wish to hatch several different breeds or varieties of chickens, or for chickens, ducks, geese, guineas, turkeys, pheasants, quails, etc.

WATER EXPANSION REGULATORS.

In order that you may get a fair understanding of the water expansion system of regulating an incubator, we will give the claim made by one of the manufacturers of that class of machines, and insert our opinion in *Italics*—said opinion being based on

93

experience of one season with one of these machines (in California), and a season with three of them (in Pennsylvania).

"REGULATION."

"This machine is regulated by the expansion of water. At one end of the tank, which contains thirty gallons of water, is attached a regulating tube some three or four inches in diameter. In this tube is inserted a float made of thin brass foil, weighing perhaps one ounce, but displacing water to the amount of one and one-half pounds. This float, with the expansive and contractile force of thirty gallons of water behind it, works with the regularity and precision of a steam engine. [*Regardless of what the heat may be in the egg chamber, which must vary according to the outside changes of temperature.*] When the water expands it raises this float, which forces up a small level bar to which is attached the extinguishers on the lamps. When this float rises, as it must do with the least expansion of water, the heat is cut off on the lamps. [*The flame is lowered, provided the lamp trip does not stick; but if the temperature of the room rises, you must be there to add more water to make it lower the flame more than usual, or to put out the light entirely.*] Should the water cool and contract, the blaze is turned on in full force. [*If the room should become cold, the operator must draw off some water, much or little, according to circumstances.*] Now, as the tank is the source of heat for the egg chamber, one can readily see that it is impossible to injure the eggs by too much heat. [*The tank being the source of heat and there being no regulator in the egg chamber to control that heater, even though you should be able to keep the water in the tank at a stationary temperature for three weeks, the temperature of the egg chamber will vary every time the temperature of the room changes—if it gets much warmer it will overheat the eggs; if much colder, they will not have enough heat. Overheating is the most dangerous.*] Indeed, so delicate is the action, that this incubator has been known to run a week without varying one degree of heat. [*When the temperature of the room has not varied.*] There are no sleepless nights connected with the use of this machine. [*We have lost lots of sleep with them (and so have many others).*] This makes the most perfect regulator ever invented, [*water must be added to or drawn from the tank according to changes of outside temperature*] giving the operator absolute control not only of the heat in the egg chamber, [*as " absolute" as with no regulator at all*] but of any given egg tray as well. As we depend upon the expansive and contractile force of the water in the tank to regulate the heat, [*when we did, and the outside temperature changed, we got left.*] of course it makes the principle which generates the superfluous heat provide for its own escape. [*If the outside temperature does not change.*] For instance, the machine running at a given point, the water being the source of heat, you cannot get any more heat in the egg chamber unless you heat the water hotter, and that is impossible unless you expand it,

[*Let the outside temperature rise 20°, 40° or 50° and you will find more heat than you want in the egg chamber, and you will not wait for the water to expand if you wish to save the eggs, but will draw off some hot and add some cold water to the tank.*] and as the expansive force of one-fourth of a degree will cut the lights entirely off, it makes it simply impossible to overheat the machine or eggs. [*It is possible at times to blow out the lamps entirely and still overheat a hot water incubator, especially during the last week, when the animal heat in the eggs is increasing rapidly.*] Manipulating the lamps does not affect the heat in the egg chamber at all. [*Rats!*] All the operator has to do is to see that he has enough of flame, and the regulator will take care that he does not get too much. [*Some operators depend entirely upon manipulating the lamps, and dispense with the regulator. If the temperature of the room would not change, the regulator might control the temperature of egg chamber; but if the temperature of the room is kept stationary, why not hatch in the room instead of buying an incubator?*] We use two lamps on our largest machines, though usually one is all sufficient to furnish all the heat required. We call one our safety lamp; for instance, should the operator forget to fill one lamp, and the light should go out, the water cools—contracts—the float is lowered, the heat is turned on in full force on the other lamp and the heat is not changed at all in the egg chamber. Suppose he forgets to attach the extinguisher to one lamp, and double heat is turned on; the water heats—expands—the other extinguisher rises and promptly puts out the other light, and still the heat remains unchanged in the egg chamber. Suppose both lights go out, for want of oil; the thirty gallons of water, packed with an inch of hair felting all around and over it, as it is, in a double cased machine, packed with the same insulating material, does not lose more than one or two degrees of heat during the whole night. [*Is a person likely to fill one lamp and forget the other, if both are together? If you forgot to attach one extinguisher on a very warm day, you would risk spoiling the eggs, as the turning down of one flame would not compensate for the one at full blast at the latter part of a hatch. A person who has not time and memory to fill the lamp or lamps once in 24 hours should not use an incubator.*] Now an incubator has got to be run three weeks, night and day. All of the above mistakes often occur through the carelessness, forgetfulness or inexperience of the attendant; so that the superiority and safety of our regulation above all others is manifest. The superiority of this principle of regulation over that on hot air machines is manifest. [*We fail to see it.*] Their only means of reducing heat is by ventilating when the heat is excessive; [*This, if applied to the heat in the egg chamber (and that is the vital place), is a mistake or a misstatement, as the best hot-air incubators do not regulate the temperature of the egg chamber by opening and closing ventilators therein or therefrom—it is the Hot Water machines that do that. The hot-air machine takes away the SUPPLY of heat from the heater and PREVENTS the heater from overheating the egg chamber at any time. See Hot-air Incubator.*] Again, other makers decry our lamp trips, saying, 'They are unreliable' [*Not only unreliable, but dangerous, from their inclination to clog and stick, causing smoke, overheat, etc.*] when this is the safest and surest part of the whole thing, [*At par with the other details.*] as it is nothing more or less than an adjustable wick tube to reduce the flame. Instead of turning down the flame, the tube is drawn up over it—the action different, but the result the same."

TWO REGULATORS.

Some incubators have two regulators. Why? Because one is not sufficient. The regulator which raises and lowers the flame of the lamp does not entirely control the heat, especially when the temperature of the room rises. To remedy this difficulty a second regulator is added, one to open and close a valve in the egg chamber. As we have used such machines, we will tell you how they worked for us, and you can decide whether or not two regulators are an advantage.

For instance we set the lamp trip regulator to lower the flame at $103\frac{1}{2}°$ and the egg chamber ventilator to open at $104\frac{1}{2}°$—both thermostats being in the egg chamber. The temperature of the room in which the incubator is operated is 55° at 7 o'clock A. M. We have filled and trimmed the lamps and turned the eggs, so we leave the incubator and go about our other business, not expecting to have to attend to the incubator again until 6 or 7 P. M. If the temperature of the room remains at 55° and the lamp trip does not clog or stick all may go well; but by 11 A. M. the temperature of the room has risen to 70°, the temperature of the egg chamber rises to $103\frac{1}{2}°$ and the lamp flame is lowered. The temperature of the room keeps rising until it reaches 80°, and notwithstanding the fact that the flame is lowered, the heat in the egg chamber rises to $104\frac{1}{2}°$ and the second regulator opens a damper valve in the egg chamber. Now the cooler air of the room (at 80°) rushes in the egg chamber, acts on the thermostat and closes the valve. It also acts on the other thermostat and causes the flame to be turned up, while the water in the tank has not cooled a degree. The higher flame now makes the water still hotter, and the air in the egg chamber is reheated, down goes the lamp, open flies the damper, the cooler air rushes in, the damper closes, up goes the flame, and hotter still gets the water in the tank. The up and down and opening and closing process goes on at an increased rate (shorter intervals) while the egg chamber goes through a course of chills and fevers with fluctuating ventilation and moisture—the latter being affected by every change of ventilation, until the temperature of the room declines towards evening to 55° or 60° or to the point at which it stood when the regulators were set or adjusted. If you can keep an even temperature in the room, you may be able to control a hot water incubator with two regulators; but in that case why not hatch in the room without an incubator?

HOCUS POCUS REGULATORS.

The old game of "hocus pocus," or, "now you see it, and now you don't," may serve very nicely as an innocent amusement for children, but when applied to the regulator of an incubator, is not particularly amusing to the operator of such incubator. The following review of a regulator farce in four acts—running through four editions of an incubator catalogue (our observations being inserted in *italics*), is to the point.

HOCUS POCUS;
OR THE MAGIC REGULATOR.

First Act (from Catalogue No. 5).

"Our Regulator." To control the heat through a course of hatching at a temperature of 103° within the egg chamber, has been the perplexing question with all incubator manufacturers. Methods consisting of electricity, mercury, water expansion, thermostat, lamp trips and other devices we have personally and practically tested, which enables us to know from experience the great advantage of an incubator with a large water capacity, to produce the amount of heat necessary without bringing it to a boil, (*Notice how they slide off of the regulator question into a large tank of water*) which is the case where small and shallow tanks are used. (*A large tank of water will not cool off as easily as a smaller one, and the possibility of overheating is greater and more dangerous. To produce a temperature of 103° in the egg chamber the body of water must be so hot; if the temperature of the room rises or falls, the water in the large tank must be made hotter or cooler, as the occasion demands, and if a large body of water gets too hot, it stays too hot longer than a smaller body of water. If it gets too cold, it takes longer to reheat it. What has that to do with the regulator?*)

The —— —— is supplied with FOUR exhaust ventilators, one at each corner of the machine, and so arranged that during incubation a continuous and evenly distributed current of warm air passes through the egg chamber, carrying with it all gas and poisonous vapor which accumulates during the process of hatching, and exhausts the heat as it reaches the point of 103°. (*With no motive power to open or close these four ventilators, will any sane person imagine that they will control the temperature of the egg chamber any more than four windows will control the temperature of a room, unless an attendant watches and opens and closes them partly or entirely as the outside temperature changes? On the contrary, the ventilation will be greater on a cold than on a warm day, the amount of air passing through being governed by the changes of temperature of the room*). By this method we have a regulator which is simple, perfect and absolutely reliable, (*The machine as described has no more regulator than an ordinary tea kettle has, and on the following page of same catalogue they not only admit that they have no automatic regulator, but you will see them say that there is no absolutely self-regulating incubator. Nothing slow about this farce!*), and one that produces a gradual and even ventilation, and avoids all chilly drafts which occur in incubators ventilated by the constant opening and closing of large swinging dampers, claimed to regulate the heat, and placed immediately over the egg drawer. We have fully demonstrated the good qualities of our hatcher, and the reliability of our regulator to the people in this whole section by completing two hatches in succession, in one of the largest and most extensively decorated show windows in the city of ———, with good success. Would respectfully ask: Could we, in any way better explain what the ——— will do? or the accuracy of our regulator. (*Have you, thus far, found any explanation as to HOW this 'regulator' regulates? How it manages the big body of water when it gets too hot and sends the temperature of the egg chamber up to 110°, as we have seen it; or how it warms it up again when it gets too cool? No? Well neither have we; but we can tell you how its manufacturers did it at the World's Fair: They turned the light up or down and waited several hours, or got others to do it for them. Had the four ventilators at the corners exhausted the heat at 103°, there would have been no danger of* overheating in the night, and had it been self-regulating, there would have been no occasion to have persons on the lookout to turn the lamp up or down.)

Many will say, after reading the description given herein: The ——— Incubator is undoubtedly a good machine, but would prefer one with a self-regulator connected with it. (*Here is an admission that it is not automatic and has NO REGULATOR, and you must regulate it entirely by hand, raising or lowering the lamp flame with each change of temperature, if you happen to get there in time.*) RIGHT HERE permit us to call your attention to the fact that there is not an incubator in existence that is absolutely self-regulating (*No, there is no perpetual motion that will fill the lamps and attend to the requirements without human aid, but there are incubators that regulate as perfectly as one could wish, and need no attention, except for a few minutes twice a day.*), it makes no difference what superior claims the proprietors set forth, or the device or arrangement used for adjusting or governing the heat in the egg chamber. (*Is not this a contradiction of their claim just made, of the accuracy of their regulator. The farce abounds with funny situations.*) Experienced incubator operators will bear us out in this statement. (*Not if they have used the best makes of modern incubators; hundreds prove to the contrary.*) From one of the many books of directions for operating Self-regulating Incubators, which we have in our possession, we quote the following paragraph. "No matter how much of a self-regulating machine it is, the SUPPLY of heat must be regulated by hand more or less, as the temperature of the room changes. You must understand the working of the regulator and see that it is set at the proper degree before you put ANY EGGS in the machine." (*No person would start to fire up a new boiler without properly adjusting the safety valve, nor would an engineer attempt to start his locomotive before getting up steam to the required pressure. A baker would not put his bread in an oven before it was thoroughly heated. Certainly an incubator manufacturer with common sense would not direct you to put eggs into an incubator before you have the heat up to 103° and the regulator set to control it. Such argument (?) is ridiculous*). The foregoing leaves you to judge the real value of self-regulators, which is materially the same advice given to all purchasers, after sending their cash for a self-regulating incubator. (*All so-called self-regulating incubators are not self-regulating, but there are some which are.*)

We most emphatically state that the —— —— is as near perfect in this respect as a machine for hatching eggs can be made, unassisted by motive power, thermostat, clock-work, electricity or any secondary appliance whatever. (*No motive power, no automatic regulation; and where there is no automatic regulation you must depend upon regulating by hand, that is by turning the flame up or down; if you are not at hand at the proper time, which may be at midday or at midnight, you get left on the hatch, your eggs get cooked or chilled.*) Our machine is in every sense practical and reliable, readily understood and easily managed. After a short experience with one of our hatchers, we will guarantee that you may be absent from the machine from seven in the morning till seven in the evening and it will take care of itself, unless (*ha!*) an unusual and extreme change of temperature takes place. A change of from seven to ten degrees will have no material effect on our incubator." (*If no extreme change of temperature takes place! Well, extreme changes do take place in twelve hours, even in daytime, but what about leaving it alone all night, when changes of from ten to forty degrees frequently occur? In some parts of this country a change of fifty degrees within 24 hours is not unusual.*)

SECOND ACT (from Catalogue No. 6, early edition.)

"OUR REGULATOR." For ten years the ——— Incubator has been manufactured. Each year it has rapidly advanced in popularity and to-day stands in the front rank with the best. This has been accomplished by adding new features—but only when such features are proven practical and consistent—with other important appliances which must be used to constitute a first-class hatcher, as an imperfect regulator will derange and seriously affect both moisture and ventilation. In equipping our ——— with regulator, we have been very careful to avoid all puzzling contrivances. (*After a constant war against regulators they conclude that they need one. That they did put their foot into a puzzling contrivance will be evident when you observe that in a later edition of catalogue No. 6 their regulator is discarded—has disappeared in twenty gallons of water.*) Methods consisting of electricity, mercury, water-expansion, lamp-trips, and numerous other devices, we have practically tested. (*Can we believe that they had tested the regulator which they now claim to have adopted, when they did not have one on their incubator at World's Fair, during the latter part of which they announced that they had a regulator, after examining the regulators of other machines on exhibition, and asking where and by whom they were made. Did they have it on the two incubators outside the fair grounds, hatching chicks for their brooders?*), but find nothing so complete in every particular as the late improved combined thermostat, which is as sensitive to the heat's action as the thermometer itself. (*How does that sound after what they said in the first act about self-regulating incubators? A month or so previous they were no good; now there is nothing so complete—and bear in mind they claim to add only such features as are proven practical. Here the farce borrows a feature of the pantomime, and the clown turns a somersault.*)

A thermostat bar twelve inches long, (*too short*) composed of steel, brass and rubber, all especially prepared for our hatchers. (*The combination is something new, indeed. We do not say that they did not attempt to use such a thermostat, but we* were told by a lady that she bought one of these incubators shortly after the World's Fair, understanding from the catalogue that it had such a regulator, but found that there was no thermostat in the machine. That she wrote to the makers about it, and they told her that the party who made them for them had failed, and they could not get any more; but that she should remember that it was not the regulator that hatched the chicks, but the incubator. She said that she returned the machine to its makers and asked for her money. We presume that she got it.) The bar is securely fastened to the under side of frame which supports the tank and on a level with the upper surface of the eggs. At the unfastened end of the bar it connects with lever and a brass rod, which opens and closes a small ventilating tube. This ventilating cap is easily adjusted by means of two little set-screws, one above and one below the cap, on outside and on top of tube, and can be set to open or close at any desired degree of temperature in the egg-chamber. With our regulator no rods or bearings are attached to the outside of the machine. (*We have not found any person who has yet seen them on the inside.*) We have known of serious accidents occurring where the regulating attachments are exposed to the meddling of children, and one instance especially, where 600 nice eggs were ruined two days before they were due to hatch by a rat running over the top of the incubator and dislocating the regulator. (*Rats! It is a case of sour grapes. Notice later, in the fourth act, cata-*

logue No. 10, where they claim to have another regulator—this time on the outside, they forget to mention the mischievous rat, or the meddling children; but such an important character as Mr. Rat should be made to play his part to the end of the farce.)

THIRD ACT (from Catalogue No. 6, later edition.)

"AS TO THE MATTER OF REGULATION." The —— Incubator has been on the market NINE consecutive years. (*Glance back to beginning of second act.*) Its growth in public favor has been rapid. It is to-day the most popular incubator in existence. It has won this high rank strictly on its merits. Its wide-spread popularity is, we claim, absolute proof of its merits—of its real value as a hatcher. The proof of the pudding is in the eating! In reply to all envious assertions made by competitors, we have this to say: Mere assertion is one thing; actual results, as described and sworn to by men and women who have bought and are now using —— hatchers, are very different things. The one is the boasting of men who have something to sell; the others are the disinterested statements of persons who have paid out their good money for artificial hatchers and have put them to the test. * * * * * The real safety of the heat governor on a successful incubator depends upon the amount of water used. That is the secret of success. The —— holds twenty gallons of water, 200 egg capacity. (*Here* *they fall back again into the twenty gallons of water. Where, oh, where is that 12-inch thermostat? Where is the new feature which they had proven; the thermostat that was as sensitive to the heat's action as the thermometer itself? Did a rat carry it off?*)

FOURTH ACT (from tenth annual catalogue, first edition).

"*AUTOMATIC HEAT REGULATORS." The ——— ———, as originally patented, was not equipped with an automatic regulator. The value of a trustworthy regulator was appreciated, however, (*What about the twenty gallons of water and the four ventilating holes, which in First Act catalogue No. 5, was a regulator simple, perfect and absolutely reliable?*) and during a number of years extensive experiments were made by us along this line. (*We should say so, when within one year four different catalogues were used to convince prospective customers that the machine had a simple, perfect and reliable regulator, while they were ringing the changes on the twenty gallons of water and four ventilators, the 12-inch rubber, brass and steel thermostat, the twenty gallons of water and four ventilators again, and then the severed band of brass and soldered steel wires.*) So-called lamp trips were tried and found to be untrustworthy. The rubber thermostat, or rubber bar was faithfully tried, but was found to lose its power after three or four hatches and thus become worthless. (*Less than a year before they said they had a complete* regulator—*the expansive power of which was, we believe, rubber. Now they say it is worthless. Then they declared that they added only such features as had "proven practical."*) A regulating device which depends on a rubber bar for its power will not last, as the rubber, when continually exposed to a temperature of 103°, dies, as the scientists say, or loses its power of regular expansion and contraction. It is like the rubber in a pair of ordinary suspenders, in the heat of summer the rubber gives out. A steel strip is used with a view to correcting this loss of power in the rubber thermostat, but with only temporary results. (*Was this true when they claimed it "proven," "complete in every particular and as sensitive to the heat's action as the thermometer itself"; or do the laws of nature change within a year to suit the convenience of these punsters? Verily, the farce is a merry one!*) We know exactly what we are talking about in this respect, for the simple reason that we once adopted this same device and had to discard it as being worse than no regulator at all. (*But say, while we are proud to see them own up for once, we can't exactly see how they are going to explain the fairy tale in one of the catalogues No. 6, where they declare that they added only such new features as have PROVEN practical and consistent, and give the public to understand that the said device had been thoroughly tested and proven, and found practical and accurate.*)

There are no secrets connected with the —— incubators, none whatever. In this book we aim to explain and describe every part and feature of our machines so that there need be no misunderstandings. We want our patrons to know beforehand precisely what they are to get. (*The bass drum and trombone put in a staccato note here.*) We are willing to leave it to their judgment whether or not our machines are honest goods built on correct principles. (*Soft cadences of harmony from second violin, flute, clarionet in E flat, French horn and violoncello. Ladies in the parquet and dress circle weep in an undertone. Lights low.*)

The regulator we now use (*For how long?*) on all our machines, depends for its power on the expansion and contraction of metals, brass and steel, under a change of temperature—a natural law that is as certain as that a stone when thrown into the air will fall back to earth. (*The old steel and brass combination was discarded years ago by parties who now manufacture good incubators, and a third tumble into the same old twenty gallons of water will probably be a feature in the next catalogue. Unfortunately the new customers very rarely see the old catalogues.*)

Tableau—Colored lights—Lively music.

CURTAIN.

There are numerous "hocus pocus" regulators (?) placed on so-called incubators. While nobody objects to would-be inventors experimenting with every new contrivance offered them as regulators, provided they do the experimenting at their own expense, before placing the machine or contrivance on the market, most persons do object to having an experimental regulator palmed off on them for a thoroughly tested and proven one. A mariner is just as safe with a deficient compass as a poultryman is with a faulty regulator on his incubator. If it occurs to you that we seem to be finding a great deal of fault, we would call your attention to the fact that when a surveyor makes a chart of a river or bay, he does not stop at lining out the safe courses and deep channels, but is equally particular to designate the dangerous rocks, treacherous shoals and sunken wrecks. Did we fail to do the same in this line you would censure us.

OTHER METHODS OF REGULATION.

Take a machine with a thermostatic bar *close to the heater*, not level with the eggs, and the bottom of the incubator being nothing but one thickness of galvanized iron, with water pans made in the iron; a change of temperature in the room will, through this thin iron, affect the temperature of the egg chamber *below* the thermostatic bar, and as high as the level of the egg centre, before affecting the bar, and under some circumstances the variations do not reach the bar at all. The chilling of the water through this thin bottom is also fatal to good results. The flame of the lamp must be raised or lowered as the temperature of the room changes to any extent.

Another incubator, partly double wall, one-inch

wood with galvanized iron inside; a thermostatic bar as in the former, or nearly so, only made in different shape and applied differently to open a damper in the top and to raise the lamp trips to lower the flames. This damper opens, and out rushes the heat—*and the moisture*. The eggs and the water are thus cooled.

This damper is supposed to open at 103½° or 104° and to close at 102°, and under certain circumstances it will do it, but you cannot depend upon it. It may open at 103½° and close at 101° for a week, and it may open at 103° and not close until the temperature falls to 95°, and again it will close at 101° and will not open until 110° is reached —we have seen them go up to 115° before opening. Can this be called "self-regulating?" Can the right amount of moisture be applied by this arrangement?

Another class have a water tank over the eggs, with thermometer immersed in the water, and the operator is directed to keep the water at a certain temperature. We have seen eggs cooked in this kind of a machine with water kept as directed.

Another lined with paper, and having a regulator four inches above the eggs, and water on top of the heater, with a damper over the water, which damper is supposed to open and close at the desired temperature. It does not do so with any regularity, and when it does open, it is liable to stay open all day and, of course, the moisture goes out. We have run three of this kind with bad results, after which they were stored away.

Another has a very pretty appearance, glass doors, etc., with thermostatic bar, single wall, clock-work and battery; dampers in the top to open and close (often six or eight times in an hour), lamp trips to lower the flame; it is quite a piece of machinery—and quite likely to get out of order, both the clock-work and the thermostatic bar, as well as the battery. When these all work right it gives very good results; but it requires skill, experience and a mechanical turn to operate this class of machines successfully.

Another has splendid arrangements for moisture and ventilation; but the thermostatic bar is affected by the swelling and shrinking of the machine and change of outside temperature; in some climates, we find it almost impossible to control it in an ordinary house.

We have found where lamp trips are used the wick becomes charred much quicker.

Another kind has a tank to pour hot water into above the eggs, which are placed in a drawer. The water must be drawn out and heated every day, sometimes several times in a day, and is not reliable for profitable work.

There are others which we might mention, but space forbids, nor is it necessary.

Most of the so-called "self-regulating" incubators that we have seen have to be governed principally by the lamp and some judgment of the operator.

Example: We have a large room made of one-inch boards, with a stove in it, which, in moderate weather with a moderate fire, will heat it comfortably. In cold weather it takes a larger fire to heat it, and in extremely cold weather the hottest fire we can make in this stove will not heat it properly. Why? Because the outside temperature penetrates the thin wall. If this same room was to be heated for chicks, the chicks to run on the floor, and you wanted the temperature on the floor at 70°, would you for a moment think of putting the regulator at the ceiling and calculating the temperature below?

THERMOSTATS.

Air, water, alcohol, ether, iodine, kerosene, mercury, gold, silver, iron, steel, brass, rubber and many other substances have been used in making thermostats, with varying success. A thermostat combining the right quality of vulcanized rubber with a grade of brass made suitable for this purpose, is the best one yet made. It took years of experiment to determine the exact quality and grade, and it requires an expert to put them together so they will work true and correctly, and neither lose power nor take a back or reverse action.

A thermostat may work near enough to control the heat of a furnace or to ventilate a house, and yet fail to give satisfaction on an incubator; for a very small variation of the thermostat may ruin the entire hatch.

Each year some manufacturer of metal goods, agricultural implements, show cases, washing machines or novelties, announces to the incubator people that he has just perfected a regulator that he will put on his incubator this season, and that he would like to furnish it at a low cost to the manufacturers of incubators. The mushroom concerns that have a regulator which does not work, and some of those that have none at all (which is far better, because you *know* you have to watch them) jump at it, and then whoop that they have something that they have tested for years. They seldom know a good thing when they see it, but the bitter complaints of customers cause them to cast about for something else, and they are ready for the next fake that comes along.

Manufacturers of thermostats for fire alarms have failed to produce one for incubators.

Air, water, alcohol, and most liquids are affected by atmospheric pressure, etc., and are not reliable. Alcohol is rarely used in thermometers now, except where extremely low temperatures are to be taken, when mercury fails to act. Zinc expands well, but fails to contract; it gradually grows longer and is useless. Mercury, if not confined, will evaporate; even when confined it is affected by moisture, and is not a perfect material for a thermostat. You may say, "If moisture affects a mercury thermostat, why will it not affect a mercury thermometer?" It does; but a comparatively small amount of mercury is used in a thermometer. As a proof of this assertion, look at the very best hygrometers, which are made by using a wet and a dry bulb thermometer. Notice the difference between the temperatures of the wet thermometer and the dry one. If moisture affects a small amount of mercury that much, what will it do with from twenty to forty times the bulk? Iron and steel are too slow to contract, or return. Gold and silver are too expensive, and are not equal to brass as a metal part. Remember that there are many kinds of brass, and all kinds will not answer. Liquids are not safe; besides the danger of a leak, the expansion and contraction of the metal in which they are confined must be overcome or compensated for.

MOISTURE GAUGES AND HYGROMETERS.

We have experimented with many moisture gauges and hygrometers both inside and outside of incubators, and have not found them of any practical use inside of an incubator. The majority of those offered to poultrymen are not at all reliable, and many of them should be classed with toys. For instance, we have two of the same make and kind, and placed them side by side, and found them indicating different degrees of humidity. We have then placed a high-priced hygrometer between them and corrected both to correspond with the high-grade instrument. In a few hours No. 1 would mark 70° while No. 2 would point to 90° while the standard instrument indicated 55°. Again, when No. 1 was at 90° No. 2 was pushing past 100°, and later when No. 2 was at 40° No. 1 was at 55°. There was no regularity or method in their variations, as one would be higher than the other one day and lower than the same the next day. Of what use would either of those instruments be in an incubator—even if you could control the moisture? The old-fashioned way of putting a cigar in the incubator would be just as serviceable. You tell by the feel of the cigar about how moist or dry the air is.

There are, of course, hygrometers that are correct, but few of them are adapted for use in an incubator. Some are too long or high for the space between the egg tray and the heat radiator, while others have a scale so small that it cannot be read without removing it from the egg chamber to a stronger light.

It is well to have a good hygrometer in the incubator room and to keep a record of its readings, daily, for it will be a valuable guide, taken in connection with the record of kinds and condition of eggs in the incubator, in determining *when* to fill the moisture pans and the amount of evaporating surface required. The time and quantity are figured out for your location and incorporated in the general directions sent by the manufacturer of a good incubator, but as he cannot know the exact kind, quality and condition of the eggs you may use, his directions are given for the best results from an *average* lot of mixed eggs and an average condition of outside temperature and humidity.

Here is the record of a little experiment, using the wet and dry bulb hygrometer and a spiral moisture gauge in the incubator room, and two spiral moisture gauges inside the incubator, the temperature of the egg chamber being 103°—the spiral instruments all being compared and set by the wet and dry bulb hygrometer, to within one-fifth of a degree, or as near as possible with such instruments: First day, in room, dry bulb 79°—wet 71°—spiral 30; inside of incubator, No. 1, 60° —No. 2, 90°. Second day, room, dry 70°—wet 76°—spiral 100°; in incubator, No. 1, 95°—No. 2, 65°; Afternoon of same day, room, dry 70°—wet 76°—spiral 85°; in incubator, No. 1, 60°—No. 2, 95°. Third day, room, dry, 75°—wet 72°—spiral, 95°; in incubator, No. 1, 100°—No. 2 index against 100°, and did not start back until No. 1 reached 75°. The same degree of heat, same ventilation and same exposure of water surface were kept in the incubator all three days. We have kept these and other tests going for years, but have given enough to illustrate the point without tiring the reader. Try them yourself. The ordinary moisture gauges are not reliable, and if they were, you could not control the outside humidity.

From one-sixth to eighteen-twenty-firsts of the incubation there is no water in the egg chamber of a majority of incubators, to create moisture. Suppose then the gauge indicates more moisture than is called for, what will you do? What use is the gauge there? If the gauge is not correct you will know no more than without one. If the water pans were full and the gauge indicated too much moisture, and you reduced it, you would probably do exactly wrong; for the gauge would fool you. There is no rule by which a given amount of moisture can be used through an entire hatch, or for a part of it, by any gauge. A correct gauge will *indicate* the degree of humidity, but thus far, it has never been perfectly controlled in an incubator, nor is it likely to be as long as ventilation is a necessity. We can regulate it to a great extent so as to make very good hatches; but the man who expects to control the moisture in the egg chamber must look beyond a moisture gauge.

There is more moisture when the chicks are hatching than at any other time; they also need more then. The tray of wet chicks increases the moisture.

Some eggs require more moisture than others. Some eggs will stand more moisture than others. Thick shell eggs require less than those with thin, soft shells. You will often hear a person say that their thin shell eggs hatched splendidly, but that the chicks died in those with thick shells. It is generally the case that they had just the right amount of moisture for the thin shell eggs, and too much for the thick ones. Would the moisture gauge help the thick shell eggs in that instance? As the majority of persons have a mixed lot of eggs, general directions must be given that will give the best results, as a rule. If directions with incubators were made to fill a dozen or twenty pages, a great many beginners would slight them and omit some of the most vital points; or they would reject them entirely as too complicated. But those who wish to get all there is in the business, should try to have eggs as nearly alike in character of shell as possible, to fill an incubator. This is impracticable to many, but comparatively easy where one has two or more incubators.

Mason's Hygrometer consists of two thermometers, as nearly as possible alike, mounted parallel upon a frame and marked respectively "wet" and "dry." The bulb of the one marked *wet* is covered with thin muslin or silk, and kept moist from a fountain which is usually attached. The principle of its action is, that unless the air is saturated with moisture, evaporation is continually going on. And as no evaporation can take place without an expenditure of heat, the temperature of the wet bulb thermometer, under the evaporation from the moistened bulb, falls until a certain point is reached, intermediate between the dew-point and the temperature of the air, as shown by the dry bulb thermometer. To find the dew-point,

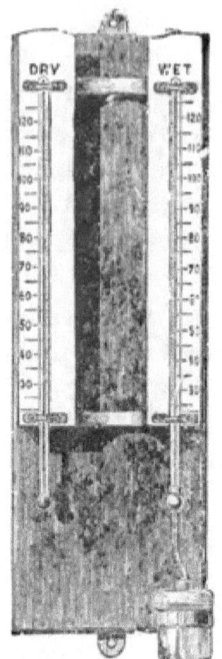

MASON'S HYGROMETER.

the absolute dryness, and the weight in grains of a cubic foot of air, tables have been constructed empirically from experiments at Greenwich, combined with Regnault's tables of Vapor Tension, for the use of which we are indebted to the courtesy of MESSRS. QUEEN & CO., *Philadelphia*.

If the air be very dry, the difference between the two thermometers will be great; if moist, less in proportion, and when fully saturated, both will be alike. For different purposes, different degrees of humidity are required, and even in household use, that hygrometrical condition of the atmosphere most beneficial to one person, may frequently be found altogether unsuitable for another. "Dry" bulb 70° and "wet" bulb 62° to 64° indicate average healthful hygrometrical conditions; any other relative condition required may easily be found by experiment, and then, dispensing with calculations, or reference to tables, it is only necessary to see that the two thermometers stand in the required relation to each other.

The price of Mason's Hygrometer ranges from $2.00 to $17.50.

TABLES FOR THE USE OF MASON'S HYGROMETER.

TABLE OF DEGREES.

Degrees of Dryness Observed.	Mason's Hygrometer — Degrees + excess x 2 = absolute Dryness.			Leslie's Hygrometer compared with Mason's.	Degrees of Dryness Observed.	Mason's Hygrometer — Degrees + excess x 2 = absolute Dryness.			Leslie's Hygrometer compared with Mason's.
	Excess of Dryness to be added.	Absolute Dryness existing.				Excess of Dryness to be added.	Absolute Dryness existing.		
0	0.0	0.0		0	11.5	1.9165	26.833		69
0.5	0.083	1.166		3	22	2.000	28.0		72
1	0.166	2.332		6	12.5	2.083	29.166		75
1.5	0.2495	3.499		9	13	2.166	30.332		78
2	0.333	4.666		12	13.5	2.2495	31.499		81
2.5	0.4165	5.833		15	14	2.333	32.666		84
3	0.500	7.0		18	14.5	2.4165	33.833		87
3.5	0.583	8.166		21	15	2.500	35.0		90
4	0.666	9.332		24	15.5	3.583	36.166		93
4.5	0.7495	10.499		27	16	2.666	37.332		96
5	0.833	11.666		30	16.5	2.7495	38.499		99
5.5	0.9165	12.833		33	17	2.833	39.666		102
6	1.000	14.0		36	17.5	2.965	40.833		105
6.5	1.083	15.166		39	18	3.000	42.0		108
7	1.166	16.332		42	18.5	3.083	43.166		111
7.5	1.2495	17.499		45	19	3.166	44.332		114
8	1.333	18.666		48	19.5	3.2495	45.499		117
8.5	1.4165	19.833		51	20	3.333	46.666		120
9	1.500	21.0		54	20.5	3.4165	47.833		123
9.5	1.583	22.166		57	21	3.500	49.0		126
10	1.666	23.332		60	21.5	3.583	50.166		129
10.5	1.7495	24.499		63	22	3.666	51.332		132
11	1.833	25.666		66	22.5	3.7495	52.499		135

By the TABLE OF DEGREES is shown, *without calculation*, the absolute dryness of the atmosphere, in degrees of Fahrenheit's Thermometer.

Observe the NUMBER OF DEGREES THE TWO THERMOMETERS DIFFER, which are here called "degrees of dryness observed," and found IN THE FIRST COLUMN of the table.

The *second column* merely contains the figures which have been added to the degrees of dryness in the *first*, and multiplied by 2, to obtain THE ANSWER PUT DOWN IN THE THIRD COLUMN.

EXAMPLE.—Temperature of the air 57, wet bulb 54 = 3 degrees of dryness observed; then add 0.5 excess of dryness = 3.5 and multiply by 2, which will give 7 degrees of absolute dryness existing.

To find the dew-point—Subtract the absolute dryness from the temperature of the air. Example 57 — 7 = 50 dew-point.

To find the actual quantity of vapor by weight in the atmosphere.—Proceed as directed in the TABLE OF QUALITY.

The comparison of Mason's with the *Dew-point Hygrometer, and of Sir John Leslie's, will be seen in the same line of the 1st, 3d and 4th columns of the Table.

TO FIND THE QUANTITY OF VAPOR BY WEIGHT EXISTING IN THE ATMOSPHERE.

PROBLEM.—The Temperature of the Atmosphere in the shade, and of the *dew-point* being given, to find the quantity of vapor in a cubic foot of air.

If the temperature of the air and the *dew-point* correspond, which is the case when both thermometers are alike, and the air consequently saturated with moisture, then in the *table of quantity* opposite to the temperature, will be found the corresponding weight of a cubic foot of vapor expressed in grains.

---

* Professor Daniel's Hygrometer is registered by the 3d column.

EXAMPLE.—Let the temperature of the air be 70° F., and the *dew-point* the same. Then opposite the temperature you have the weight of a cubic foot of vapor—8.392 grains.

But if the temperature of the air be different from the *dew-point*, a correction is necessary to find the exact weight.

EXAMPLE.—Suppose the *dew-point* be 70° F., as before, but the temperature of the air in the shade be 80°, then the vapor has suffered an expansion due to an excess of 10°, which requires a correction.

We find in the *table of corrections* for 10° is 1.0208.

Then divide 8.392 grains at the *dew-point, viz.*, 70° by the correction, corresponding to the degrees of absolute dryness, *viz.*, 10°, and you have the actual weight of vapor existing.

EXAMPLE. $\dfrac{8.3920}{1.0208}$ 8.221 grains existing, which substracted from the weight of vapor, corresponding to the temperature of 80° F., gives the number of grains required for saturation at that temperature.

EXAMPLE.—11.333 gr. at the temp. of 80° F.
      8.221 gr. contained in the air.
      ─────
      3.112 gr. required for saturation.

To find the relations of these conditions on the natural scale of humidity [complete saturation being 1.000], divide the weight of vapor at the *dew point* by the weight at the temperature of the air, the quotient gives the parts of 1.000 the degrees of saturation.

EXAMPLE. $\dfrac{8.392 \text{ gr. at the } \textit{dew-point}—70°}{11.333 \text{ gr. at the temp. of the air } 80°} =$

7.40 degrees of humidity, saturation being 1.000.

The principles of these calculations will be found in Professor Daniel's *Meteorological Essays;* in Mr. Anderson's Essays on Hygrometry; in the Edinburgh *Encyclopædia*, Vol. XI, and in the Edinburgh *Journal of Science*, Vol. VII, page 43, in an excellent article on the Dew-point Hygrometer, by Mr. Foggo, from which the table of corrections has been partly subtracted. The table of quantity by weight has been taken from Professor Daniel's Work on Meteorology, to which the reader is referred for further particulars.

TABLE OF QUANTITY.

SHOWING THE WEIGHT, IN GRAINS, OF A CUBIC FOOT OF VAPOR, AT DIFFERENT TEMPERATURES, FROM 0 TO 95° F.

Temp.	Weight in Grains.	Temp.	Weight in Grains.	Temp.	Weight in Grains.	Temp.	Weight in Grains.
0	0.856	24	1.961	48	4.279	72	8.924
1	0.892	25	2.028	49	4.407	73	9.199
2	0.928	26	2.096	50	4.535	74	9.484
3	0.963	27	2.163	51	4.684	75	9.780
4	0.999	28	2.229	52	4.832	76	10.107
5	1.034	29	2.295	53	5.003	77	10.387
6	1.069	30	2.361	54	5.173	78	10.699
7	1.104	31	2.451	55	5.342	79	11.016
8	1.139	32	2.539	56	5.511	80	11.333
9	1.173	33	2.630	57	5.679	81	11.665
10	1.208	34	2.717	58	5.868	82	12.005
11	1.254	35	2.805	59	6.046	83	12.354
12	1.308	36	2.892	60	6.222	84	12.713
13	1.359	37	2.979	61	6.399	85	13.081
14	1.405	38	3.066	62	6.575	86	13.458
15	1.451	39	3.153	63	6.794	87	13.877
16	1.497	40	3.239	64	7.013	88	14.230
17	1.541	41	3.371	65	7.230	89	14.613
18	1.586	42	3.502	66	7.447	90	15.005
19	1.631	43	3.633	67	7.662	91	15.432
20	1.688	44	3.763	68	7.899	92	15.786
21	1.757	45	3.893	69	8.135	93	16.186
22	1.825	46	4.022	70	8.392	94	16.593
23	1.893	47	4.151	71	8.658	95	17.009

TABLE OF CORRECTIONS.

TO BE USED WHEN THE TERM OF DEPOSITION, OR DEW-POINT, DIFFERS FROM THE TEMPERATURE OF THE AIR IN THE SHADE.

Diff. of Temp.	Correction.	Diff. of Temp.	Correction.	Diff. of Temp.	Correction.	Diff. of Temp.	Correction.
0	0.0000	13	1.0271	26	1.0542	39	1.0813
1	1.0020	14	1.0291	27	1.0562	40	1.0834
2	1.0041	15	1.0312	28	1.0583	41	1.0854
3	1.0062	16	1.0333	29	1.0604	42	1.0875
4	1.0083	17	1.0354	30	1.0625	43	1.0896
5	1.0104	18	1.0375	31	1.0646	44	1.0917
6	1.0125	19	1.0396	32	1.0667	45	1.0937
7	1.0146	20	1.0417	33	1.0687	46	1.0958
8	1.0167	21	1.0437	34	1.0708	47	1.0979
9	1.0187	22	1.0458	35	1.0729	48	1.1000
10	1.0208	23	1.0479	36	1.0750	49	1.1021
11	1.0229	24	1.0500	37	1.0771	50	1.1042
12	1.0250	25	1.0521	38	1.0792	51	1.1062
...	....	...	....	...	....	52	1.1083

RULE.—To find the weight of moisture in a cubic foot of air at any time. Divide the *weight in grains* found opposite to the *temperature*, corresponding to the dew-point at the time, in the *table of quantity*, by the *correction* found opposite to *difference of temperature* in the table of corrections, corresponding to the absolute dryness existing at the time.

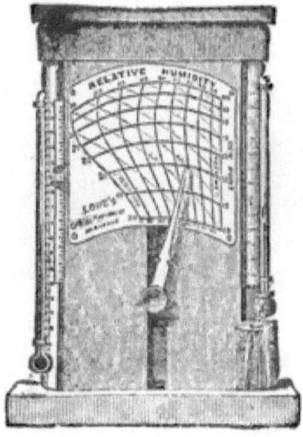

MASON'S PATENT HYGRODIKE.

This instrument is on the principle of Mason's Hygrometer, but arranged with dial and pointer so that the absolute and relative dryness and the dew-point may be read off without calculation. The price is $15.00

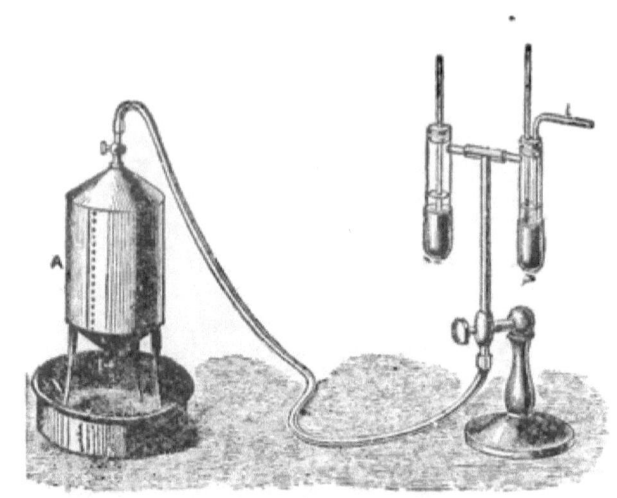

REGNAULT'S HYGROMETER, WITH ASPIRATOR.

These instruments consist of a thin and highly-polished tubular vessel of silver, having one end somewhat longer than the other. A very delicate thermometer is introduced into the tube at the smaller end, to which end of the tubular vessel, also, a flexible rubber tube with ivory mouth-piece is attached. A sufficient quantity of ether to cover the bulb of the thermometer, being poured into the silver vessel, the ether is agitated by breathing through the flexible tube. A rapid evaporation ensues until at the moment the dew-point is reached, the moisture is seen to condense upon the exterior surface of the polished silver tube. The reading of the thermometer at this precise moment gives the dew-point. Complete in case, $75.00. Too high priced for ordinary use, but a splendid addition to an experimental outfit.

BROODING.

For the first day after the chicks are taken from the incubator they should be confined to a brooder, to get them used to it, so they will go in and out.

The brooder or nursery should be so constructed that they can go in and out at will, and not be compelled to stay under the hover when too warm, or outside when cold, as is the case with too many brooders, the chicks in the middle being made prisoners by those on the outer edges and injured or suffocated, while at the same time those near the outer edges are perhaps suffering with cold.

The illustration shows a very nice arrangement for a nursery or indoor brooder, and is convenient to have even when you have large brooding houses heated by hot water pipes. This brooder can be made by anyone who is handy with tools—the metal heater and chimney can be made by any tinsmith.

A brooder is supposed to take the place of a good hen. To do this successfully it must be made as nearly like a hen as possible. Now how is a hen built? Where does the heat come from? Where do the chicks hover? How do they get to and from the heat, and receive fresh air? Look at the illustration of a brooding hen, and see for yourself. Is not the heat which the chicks get from her principally side heat? By chance a chick may get caught under the breast bone or under the foot of a hen, but not often. The wings, feathers and down of the hen retain the greater part of the heat from the body. The brooding chicks can put their heads out for fresh air, instead of being crammed into a bunch and surrounded by from fifty to a hundred other chicks. If they are too warm they can get out, if not pinned down under the breast bone or foot of the hen.

The heat from the hen certainly cannot be termed "bottom heat," nor yet "top heat." It is—as she squats down and her body is *surrounded* by the chicks—principally "side heat," with some top heat retained by her feathers. Nature provides a covering for the chicks to nestle under, and a brooder should have something soft and heat retaining for them to huddle under; not simply a top, compelling them to squat down on the floor, but something to take the place of the feathers of the hen.

For the first week the temperature of the brooder should between 80° and 90° at about two inches above the floor. After a few days' practice one easily learns to test the heat under the hover of brooder by the feel of the hand, and the thermometer is then unnecessary. When we say that the temperature should be between 80° and 90°, we mean that it should be that warm under the hover without any chicks under it. As the chicks grow older they require less artificial heat, because they furnish more animal heat as they increase in size, so you should gradually decrease the heat; but never have it lower than 70° *with* chicks under the hover, as long as they require brooding. This minimum degree of heat should be reached at about the sixth week.

It is not necessary to use the thermometer after once adjusting the heat supply of a brooder, because (with a proper brooder) you can tell by the action of the chicks if they are too warm or too cold. If too warm they will put their heads out from under the hover or come out entirely. If too cold they will chirp in a tone which no one can mistake for a signal of satisfaction. Learn to tell the right temperature by placing your hand under the hover—it is very simple and easy, and less trouble and more satisfactory than a thermometer. Some persons will advise you to have a brooder too hot rather than too cold. We say have it just right. If the brooder feels comfortably warm to the hand, and the chicks stay under the hover (at hovering time), seem contented and do not cry out, you may be sure they are all right.

Brooders with either "top heat" or "bottom heat," and having a square or oblong hover, say, eighteen to twenty-four inches square or wide, either with or without flannel or woolen drapery, are open to very serious objections. When the chicks in the middle get too warm they try to move to the outer edges or to get outside entirely, but those on the outer edges, being comfortable or just a trifle cool, refuse to stir, and the ones in the centre must remain there and become overheated and sick. Or, if the chicks on the outer edges become cold they crowd toward the centre and crush or smother the chicks that are there. You say, why not have the heat just right, so they will have no occasion to push or crowd? How can you, with that kind of a brooder? Place a crowd of people under a shed the floor of which is heated, and will not those in the middle be uncomfortably warm if those on the outside edges are just warm enough? Or, if those in the middle are just warm enough, will not the outer ones be cold, particularly if the weather is extra warm or extremely cold? It would be the same if the heat came from the top. But let chicks *surround* either a square or oblong heat reservoir or heater, so arranged that the hover projects only far enough to shelter two or three rows of chicks (only two or three deep from the outer row of flannel drapery to the wall of the heat reservoir), and crowding is impossible; the inner chicks can get out or the outer ones can get closer in. If a chick is pushed from under the hover, another takes its place, and the ousted chick finds the vacant spot and occupies it. The inner row of chicks are only about six inches from the outside air and do not suffer for want of pure ventilation.

You can place a hundred men in rows two or four abreast and they will be comfortable, but place them in a square or round room just large enough to hold them, and, no matter whether it be winter or summer, at least one-third of them will be uncomfortable. Is not the argument conclusive? Still we do not give this from theory, but from experience—after burying bushels of chicks from both "bottom heat" and "top heat" brooders.

Since adopting "side heat" we have not lost over five per cent.

The flannel or woolen drapery which hangs down from the hover and helps retain the heat and gives a feeling of cosy comfort to the chicks is essential. Nature gives them side heat (from the hen) and soft covering (the feathers of the hen), and so must we if we want them to be comfortable and thrifty. Heated floor or ceiling is not enough. Would you like to heat a bedroom up to 70° or 80° and lie on the bed or floor with no covering? We think you would prefer to have the room at 30° or 40° and put on a few blankets. Use your best judgment in the matter of brooding. If your present system is not satisfactory, or if you have not begun, try the side heat, which combines partial top heat, as shown by the illustration of the brooding hen, with narrow hover well draped with something to take the place of feathers, and you will solve the problem of brooding. This plan takes from one-third to one-half less fuel than other styles of brooders.

After the chicks have been in the brooder or nursery one day and night they should be allowed more run, and if the weather is fair they should have out-doors runs. Keep them from the grass until the sun or wind has dried it. If the weather is cold, watch them the first day and see that they do not stay out and get chilled; but after they have learned to go in and out of the brooder, they may be let out in winter as well as in warmer seasons, but you must use some judgment.

Keep the brooder clean by using sand or earth on floor of brooder and house. Do not use the very fine, dusty kind of sand if you can get anything else.

As soon as the chicks show an inclination to roost get them out of the brooding house and into less expensive houses, if you have them, and make room for others, besides giving them more range.

BROODING HOUSES.

A continuous brooding house, divided into rooms, should have a passage way through it, and if thirty feet long, or longer, should have a hot water stove to furnish heat for the brooders, so as to save attention to so many lamps.

If the house is furnished with single brooders, each room should be five and one-half feet wide and nine feet long—the length being parallel with the divisions or dividing fences of yards. If heated by continuous hot water pipes, the rooms should be nine feet wide—at right angles with the dividing fences, and five and one-half feet long—parallel with dividing fences. The reason for this arrangement will be apparent on examining the different brooding apparatus. These rooms will accommodate from fifty to one hundred chicks, each.

Von Culin's Brooding House for Hot Air Brooders

138

BROODING HOUSE
WITH HOT AIR BROODERS.

This brooding house is fitted with the Von Culin Indoor Hot Air Brooders. The rooms may be five feet wide and nine feet long, or may be made wider, if desired. The position of each brooder is shown (B). The floor is elevated one foot above the ground, to allow brooder floor to be on a level with the room floor. The top frame shown around the hot air brooder is moveable, and may be taken away after the chicks get used to the brooder. Or, it may be covered with fine wire netting and placed on at night where there is danger from rats or other vermin.

BROODING HOUSE
WITH HOT WATER SYSTEM.

This brooding house is fitted with the Von Culin system of hot water piping and brooders. The drawing shows the arrangement of the pipes and brooders, the division of rooms and yards, the arrangement of base board and wire divisions, and the passage way back of the rooms. The second drawing gives details of the brooding system.

Each room should be nine feet wide; this gives a brooder nine feet long. The depth or width of

Von Culin's
≕ Hot Water ≕
Brooding House,
Side Heat System

rooms, five feet, with yards as large as you can make them. The base boards should be two feet high, and on top of that four feet of wire netting,

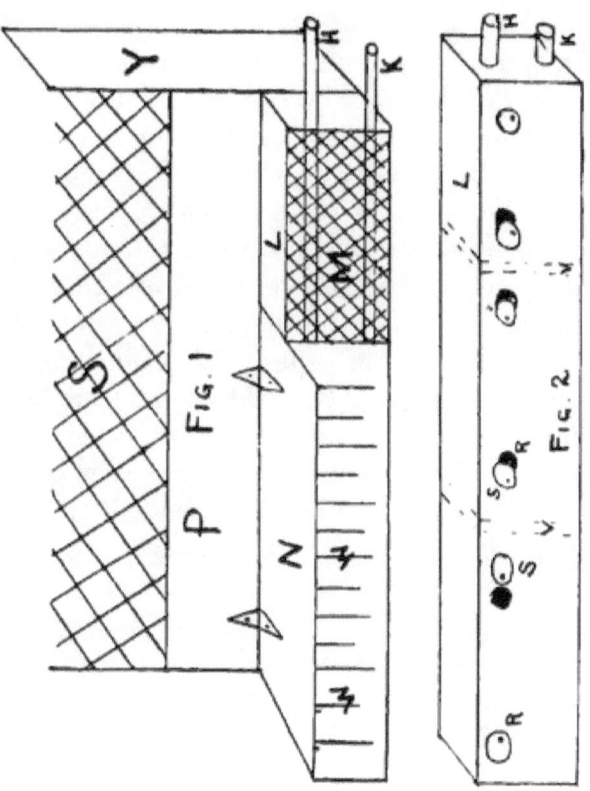

VON CULIN'S BROODING SYSTEM.
(For Hot Water Circulation.)

two inch mesh. If base board is only one foot high, then use one inch mesh. Posts should be ten feet apart. Passage, two and one-half feet wide. It does not cost any more to build rooms this size and shape than to build them long and narrow. The wide room gives a better shaped yard.

The preceding illustration represents the Von Culin system of brooders, heated by circulation of hot water for houses of all sizes. The illustration on page 140, shows the application of the same. Y, shows the passage of house; D, the floor; P, the base board above the brooder; S, wire netting which divides the passage from rooms; H, the main hot water pipe; K, return pipe; L, the boxing of pipes, forming the heater; M, wire netting to keep chicks from 'pipes; W, flannel covering the wire, and strips of flannel hanging from hover; N, hover, which is hinged, and can be raised for cleaning; V, divisions to separate each brooder heater; R, S, are two one and one-half-inch holes opening from heater box into passage, and having a round button, to regulate the heat for each separate brooder, to suit chicks of any age. All pipes are run on a dead level; and all hovers the same height, though the hover can be lowered to any desired degree to accommodate chicks of any size. It is not necessary to be continually shifting chicks from one room to another; the height and temperature of each or any hover may be changed at will.

HOT WATER STOVE
WITH CONTROLLING APPARATUS.

This Hot Water Stove, which has a water jacket around it, supplies the heat to the hot water pipes of brooding system, or for other purposes; the water flowing out through the upper pipe and returning through the lower one to be reheated, keeping up a continual circulation. The tank above is to supply any escape of water in the form of steam which may occasionally be generated by overheating. The safety valve prevents a pressure of over five pounds, as pressure above that point would stop the circulation of water. Each day the valve below the tank is opened to supply any waste which may have occurred, and then closed again, and kept closed while heating the brooding house. If hot water is needed, to mix feed, scald chickens, etc., the two valves from the brooder pipes are closed and the valves to and from the tank are opened, thus causing a circulation through the tank and boiling the water. To turn the circulation through the brooders again, you simply close the valves nearest the tank and open those leading to and from the brooders. With the Von Culin system two pipes only are used in the brooders; other pipes may be run into the rooms or passage. With the "top heat" system four pipes are used. Either system is simple, and may be laid by any handy mechanic or plumber. Many poultrymen make and fit up all their brooders and houses.

SINGLE OUT-DOOR BROODERS.

Where the climate will permit and you have plenty of ground, the outdoor brooders can be used without division fences, and in most places, without any permanent fence at all, by simply having one or two portable fences to place around the brooder for two or three days, to colonize the brood, then taken away and used for other brooders. Scatter the brooders around the field or lawn, and the chicks will have grass and larger range than otherwise, which always reduces the expense of raising them, gives them quicker growth and better health. It is a profitable way to raise broilers.

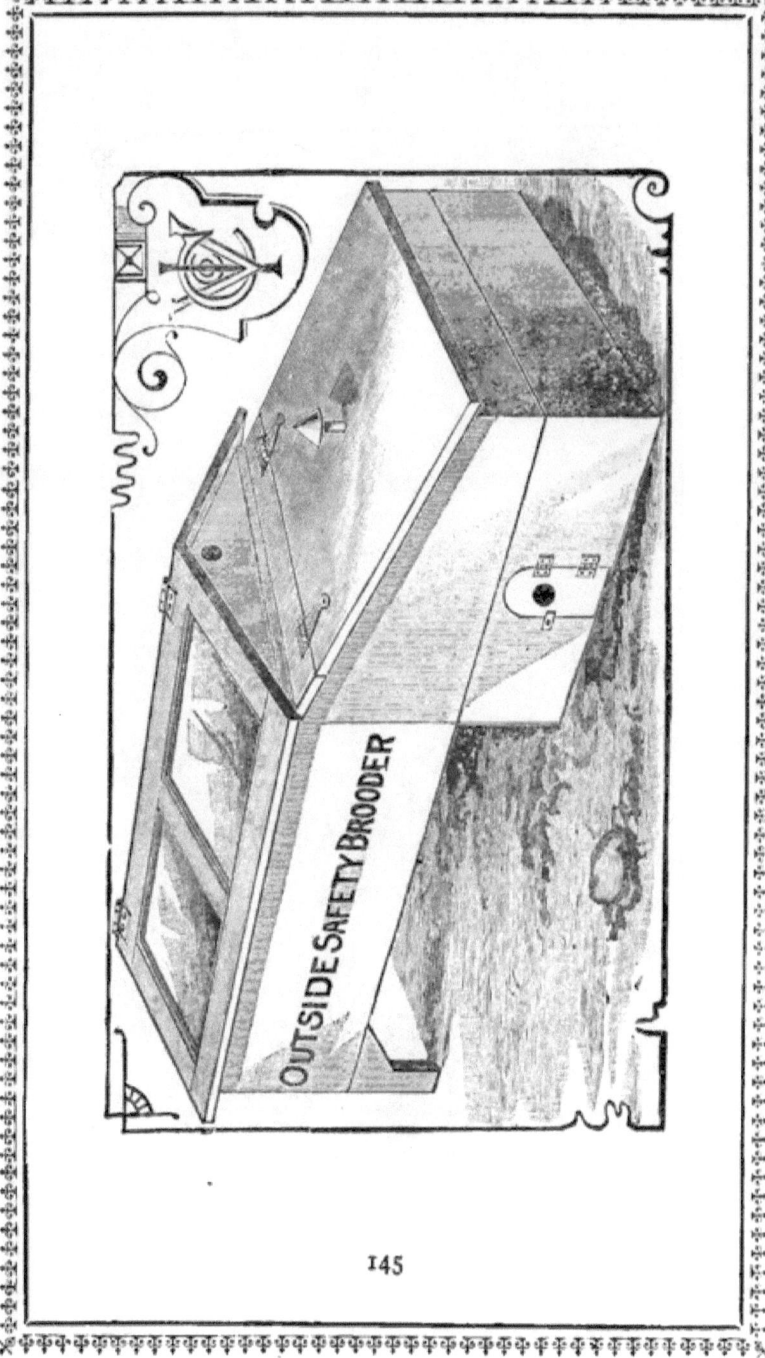

OUT-DOOR HOT WATER BROODER.

This illustration shows a brooder with six compartments, each of which holds one hundred chicks. The brooders are made of various sizes, to hold two hundred, four hundred or six hundred chicks, and are heated by a lamp. The drop runs are shown on one side, while those on the other side are up. The runs are kept up the first-day, until the chicks get used to the brooder, then lowered.

Moveable runs are made to keep the broods separated. They can be made wide at the farther ends by placing the corner yards at an angle. If these runs are made near a division fence some of the larger chicks may be let out into a larger yard or a free run at your convenience. These brooders need no houses, but may be placed in a field, orchard, or yard. In severe climates they may be put under a shed in the winter, and moved out later in the season. They are principally side heat with moderate top heat. It is impossible to overheat, or for the chicks in them to crowd and smother. It is the same principle as used in our Hot-water System in Brooding Houses.

The brooder which has a number of compartments should have a yard for each compartment, or for each brood; but should be made movable, so that the whole plot or group of runs can be plowed or spaded, or the entire arrangement moved to new ground.

BROODER YARDS.

All brooder houses should have yards or runs, the larger the better. The fence should consist of a base board one or two (two preferred) feet high, nailed to posts, two by four inches, placed ten feet apart and two feet in the ground and six feet above ground, with wire netting above it. If the base board is two feet high the mesh of the netting may be one and a half inch or two inches, but if the base board is only one foot high, then you should have one inch mesh netting at least one foot above the board, and finish above with larger mesh. The same kind of fence should divide the yards or runs. If possible have an independent gate at the outer end of each yard. You will soon recognize its usefulness. Besides keeping the chicks separated and at peace with neighbors, the base board breaks the wind in winter.

FEEDING CHICKS.

Give them no food for the first twenty-four hours, as the yelk of the egg is absorbed by the chick just before it breaks from the shell, and supplies nourishment for at least twenty-four hours after hatching. Cramming other food into the stomach before the yelk is digested is injurious.

Give them fresh, clean water from the start. Have it in a fountain so they cannot get into it.

If water is given them in open vessels they will stand in it, splash themselves and get chilled, which is dangerous in cold weather and undesirable at all times. They will also make the water filthy and unfit to drink.

We have tried the plan of giving no water for the first three weeks, with very good results; but the trouble which begins when water is first given to them makes the method undesirable and very risky, as they will often drink until they fall over or froth at the mouth.

For the first week give millet seed, as much as they will eat up clean, *every two hours*. If you cannot get millet seed (which is sold by all seedsmen), or will not use it because you think it is too high, give finely cracked wheat, corn, or the grain which is raised in your section. Sieve out the coarser parts for larger chicks as it would be wasted if fed to chicks under a week old. Place granulated charcoal and grit where they can get it at will. Grass is good for them at all times.

Second week, give millet seed, fine cracked wheat, cracked corn, and occasionally (about twice a week) rolled oats. If millet seed is high in price, discontinue it the second week and keep on with the cracked grain. Let the cracked grain be fine through the day and coarser at the last feeding in the evening. Do not mix the grain, but give each kind separately—not two kinds at any meal. The change gives them a relish. Once a day give them a feed of finely ground green bone (fresh), about the bulk of a grain of corn for each chick. If the chicks are confined give finely chopped onion tops or onions. Feed every two hours.

Third week, give the same food as second week, but increase the quantity of green bone one-half. Feed five times a day.

Fourth week, still feed five times a day, and to the bill of fare add cooked fresh meat, finely ground, once a day, the quantity for each chick being the bulk of three grains of corn.

Fifth week, morning food, first day, two parts ground corn, two parts ground oats, one part bran. Mix with just enough water to make it stick together (do not have it sloppy). Use hot water in cold weather. Always allow it to stand fifteen minutes after mixing, to swell. Have the corn and oats ground together, they grind better that way, and make a better mixture. Feed cracked grain the balance of the day, that is at noon, in the middle of the afternoon and about a half-hour before they go to the hover for the night, making four meals a day. Second day: soft feed as above for first feed and at noon. Cracked grain balance of the day. Third day: three meals of same soft food, and cracked grain at last feeding. Same for the balance of the week.

Sixth week: same soft food at each meal except last one at night, increasing the allowance of chopped, cooked meat. Meat should be given between meals, but never give more than the bulk of a chestnut to each chick.

Seventh week and until marketed, same as sixth

week, and give all they will eat up clean at each meal, except meat.

Grass is wholesome for chicks as soon as they begin to eat. If they do not have access to grassy runs or yards, it is well to cut a few fresh sods as often as convenient, and place them in the runs. If this cannot be done, fresh cut grass is good. When grass is out of season, finely chopped cabbage once a day or every other day to chicks over three weeks old.

Do not forget the *charcoal*, grit, and a box of crushed shell for each brood.

FATTENING BROILERS.

Fattening broilers by close confinement is a mistake. Try to put on all possible flesh by giving them all the food they will eat up clean, and the more exercise they have the better their appetite will be, the faster they will grow, and the hardier birds they will make.

If you undertake to force chicks under four weeks old by soft food, you will impair their digestion, cause them to be weak in the legs, and to feather fast. You may gain a little flesh by the soft food from the start, on those chicks which do thrive; but that will be overbalanced by losses from leg weakness, diarrhœa, forced feathers, etc. By any method of feeding some chickens will be fatter and plumper than others. It is ridiculous to talk of fattening pen broilers.

Would any sane man undertake to fatten a young pig if he wanted it to grow and put on flesh? Would you fatten or try to fatten any stock that you wished to grow?

You must make bone and muscle first, then put on flesh. To do this you must keep your chicks (or other stock) in good health. By overloading man, beast or fowl with unnatural food you are almost sure to disarrange the system; and soft, sloppy food is not the natural food of chicks or fowls. If you are determined to give soft food to chicks under four weeks old, bake corn cake in the oven, and make it so that it will crumble.

Never feed boiled eggs to chicks.

OLD FOWLS AND YOUNG CHICKS.

Keep old fowls away from the brooders and brooding houses and runs where incubator chicks are kept, and do not mix the chicks which were hatched under hens with those hatched in incubators, because the chances are nine to one that lice or mites will be communicated to the latter.

Chicks hatched in incubators are (cleanliness having been observed) free from vermin; but we have known a whole section of brooder houses to be filled with lice by placing a single brood of eleven chicks, whose mother died, in a brooder with other chicks. The new comers were not suspected of being lousy, and the lice multiplied and spread through all the adjoining rooms and yards before being discovered.

We have also known roup to be spread in the same manner—in one instance breaking up the establishment. In the latter instance the proprietor was warned, but he knew it all, and had it his own way.

Things which seem small or trifling sometimes make tremendous results. By watching and directing small matters we control greater ones.

SELECTING BREEDING STOCK.

Where a large number of fowls are to be bought at one time it is not easy to get just what is most desirable, but care should be used that no objectionable fowls are bought or kept.

If you are starting a new plant, you want young stock. *Do not start with a lot of old hens,* for they will certainly give you a set back that will not only dampen your ardor, but kill your profit the first season, and perhaps cause you to make a total failure.

The old notion that an old hen would produce better and stronger chicks than a young hen, has died a natural death, and is laughed at by the majority of experienced poultrymen. If you want hens to hatch with, the old ones are all right; but for ordinary use they have the following disqualifications: they lay fewer eggs than younger birds, they are more liable to disease or return of previous ailments, very liable to become overfat, have often acquired bad habits, such as feather pulling, egg eating, laziness; and when killed and dressed for market do not please the customers or bring new trade.

By all means select young stock. You will be compelled to keep over for the second season about one-half of the stock you start with, if you intend to establish a good system on a large scale, and you will then find that you have plenty stock that is as old as you want it.

Ordinarily you can tell the young stock by examining their legs, heads, combs and plumage: the legs being smooth and clean, the heads bright and clear, the comb smooth and not too large; there will be an absence of the short spurs which are found on some old hens. The plumage will be fresh in color, without the brassy or dull appearance shown in the plumage of old stock. In old fowls of the white varieties the yellow tinge shows plainly in the males, yet some of the young cocks show it plainly. The novice will find no difficulty in telling old from young males; but with pullets it is not always so easy. Some pullets have very rough legs, and would be taken for hens two or three years old; but they are not desirable, so you need not make that mistake. On the other hand you will often find hens four or five years old with as clean legs as pullets have. A young hen or pullet does not generally have as rough a comb or wattles, and is more active.

CULLING BREEDING STOCK.

Cull your stock as often as you can find any culls. Culls of any kind are undesirable. If you find a hen that is a poor layer, get rid of her and replace her with a better one. She will generally eat as much as a good layer, takes up as much room, and requires as much care and attention, except in gathering eggs.

An egg-eating or a feather-pulling hen, no matter how fine a bird or how good a layer, should be killed and marketed or eaten. In a very short time she would teach the other fowls her bad habits. An egg eater will sometimes eat three times as many eggs as she lays.

The hen that wants to sit quite often may be useful in that line, but no other, and should be got rid of.

The lazy hen that lingers on the roost late in the day is not the one that lays the eggs. Cull out all of that kind, and either replace them or keep fewer hens; they ea the profit made by good hens, and are a serious drawback.

WHEN TO CULL.

First cull before you buy, next as soon after as you can determine what to cull. Then cull the chicks at broiler age, keeping the best in form and color for breeding or for eggs, then cull as often as you find anything to cull.

THE BUSINESS HEN.

The business hen is the one that brings the profit. As a rule, you will find that the lively, quick moving hen or pullet is a good layer. A large, bright comb also indicates a good layer. Where a limited number of fowls are kept, the best and the poorest layers may be discovered by the shape, color and other peculiarities of the eggs, and in such cases the culling and selecting can be carried to much finer points than usual.

A SECRET.

We call the following a secret because few persons outside the charmed circle of successful poultry culture know it, and only the *leaders* of the successful make full use of the knowledge of it.

It is *the* secret of success with poultry on a large scale, and is practised by the most successful poultrymen throughout the world.

Kill and burn all diseased chickens and fowls as soon as discovered.

Observe the flocks *every* day and visit their houses *every* night. If you hear any wheezing or sneezing, or see any shaking of heads, use the vaporizer described elsewhere *promptly*.

Feed no carrion or tainted meat nor any spoiled, musty or damaged food of any kind. See that the meat you use is sweet and fresh, the grain good and sound ; the water fresh and clean, cool in summer and not frozen in winter. Supply plenty of gravel or grit, lime in some shape—crushed shells or otherwise, and granulated or broken charcoal all the time.

Give your birds clean and comfortable quarters.

Just as soon as a hen ceases to be profitable market it. When you see a fowl that does not look like a business fowl—one that stays late on the roost, stands around by itself, is inactive, overfat, broken down, walks unsteadily, is pale, or shows signs of moulting, dress and market it *promptly*, except where you wish to keep moulting hens over for the next season, and they are not the most profitable kind to keep.

Remember that this rule does not apply to diseased fowls. The first part of the "secret" disposes of them.

When these rules are strictly followed there will be no danger of diseased fowls. You may say that you do not wish to sell or reduce your stock. Perhaps not, but the birds we have described, the overfat, inactive and moulting fowls are fit subjects for disease, which in every form is more easily avoided than cured, and your risk of loss is far greater with a few such fowls among your flocks than from a reduction in numbers by marketing the same.

A VILLAINOUS PRACTICE.

The preceding rules, if strictly carried out, prevent and remove all temptation toward the villainous practice of marketing diseased fowls. It is almost incredible, yet a fact, that many extensive breeders and commission men kill, dress and market diseased poultry. In a State which we need not name, we have seen fowls in all stages of roup, chicken-pox, canker, etc., offered for sale by the poultry dealers and commission men, for table use, in the same markets and stores with fish, meat and game. Hens so badly diseased that they could not eat, and those with heads swollen to almost double their normal size, were dressed *heads off.* Never sell a fowl that you cannot leave the head on. Never buy one with the head off.

Of course you indignantly disclaim any idea of selling diseased fowls, and we are inclined to believe that you would not do so intentionally; but unless diseased fowls are killed and burned as soon as discovered and the utmost vigilance exercised to discover them, you are very apt to be the unintentional assistant in such business. For instance you have a lot of fowls to sell, in order to make room for growing stock. A buyer or huckster comes along and makes you an offer for the entire lot. You have not been diligent in examining them of late, and a mild form of some disease has got among them without your notice, or you have a few quarantined, and you let them go just as they are, because, in the latter event, he tells you that he thinks he can cure them ; and they all go into the coop together. When he gets them to his killing house he cures the worst ones first, with knife or hatchet, whichever way looks best. Some eggs look best scrambled; some fowls look best "heads off."

Men who deliberately do such things would be more useful to the poultry fraternity with heads off.

Poultrymen, as a rule, are first-rate people, but there are poultry jockeys as well as horse jockeys. The horse trader who doses and doctors a nag to conceal a dozen imperfections from the innocent purchaser, is a saint in comparison with the man who knowingly sells a diseased fowl for food.

Of what interest is this to the honorable beginner? It shows him an existing, hidden danger and the means by which to avoid it. If he is on his guard against selling diseased fowls, he will also be on his guard against buying them, and that one point may decide his success or failure in *raising* poultry. One diseased fowl may spread ruin throughout the establishment.

THE VAPORIZER AND ITS USE.

One of the most useful implements on the poultry farm is a vaporizer. If you have not got one, you can easily make one. Take an ordinary hand lamp which has a No. 1 burner, and make a tin chimney for it, similar to an incubator or brooder chimney, with mica piece in front to show flame. Take a *seamless* tin box, about six inches in diameter and any convenient depth, having a cover to fit, and, with strips of tin and rivets, fasten the box to the top of the chimney—over the top and about an inch above it, so as to leave a good draft for lamp. If you cannot find a box of convenient size, get a small cake pan or saucepan, and fit a flat tin lid to it. Having your vaporizer complete, keep on hand (in air-tight jar or box) a supply of carbonate of ammonia and gum camphor.

When you discover any colds, wheezing or sneezing among your flocks, mark the house or the houses, and after dark cover all the cracks and close the doors and windows, and having put a convenient quantity of carbonate of ammonia and green camphor in the tin box (two ounces of carbonate of ammonia to one ounce of green camphor), light the lamp, and burn it in the poultry house. The two ingredients being volatile, will vaporize and fill all parts of the house. Burn until the fowls move about on the perches and show signs of uneasiness, and the vapor gives a strong, pungent odor. If the birds attempt to leave the roosts, remove the vaporizer. For an ordinary cold one fumigation will generally effect a cure, and the fowls need not be removed from their house; it is also a good preventive to be used occasionally when colds or roup are prevalent in your neighborhood. Cases of roup, diphtheria, canker, etc., should be removed to a quarantine and treated there. Our advice is to kill and burn them at once; but as we know that some *will not* do this, then we must advise what is next best. Having placed these cases in quarantine, close the roosting houses and use the vaporizer each night for at least three nights. If in the early stages, three treatments will cure, but later on it requires more. We have repeatedly cured bad cases of roup and diphtheria with the vaporizer, and have never failed when taken in time. When not in use always keep the box containing the ammonia and camphor closed. The success of this treatment depends upon using it *promptly*, and in bad cases, perseveringly. Those who have had experience in treating roup by old methods will gladly drop the disgusting, sickening processes, and substitute this clean, wholesome, and convenient one. It is a good plan to keep lantern and vaporizer close together, and when you make your night round with lantern, just carry the vaporizer along.

EGG AND BROILER FARM.

The above engraving shows a section of an egg and broiler farm built by C. Von Culin in 1885, reproduced from a photograph. In the foreground are a group of sixteen brooding houses under four roofs (four under each roof), with yards attached. Beyond are laying and roosting houses scattered at regular intervals. They are built on runners, and may be moved to new ground as often as required for cleanliness and new pasture. There are no fences around these houses; a small, portable fence is placed around a house for three days to colonize a new flock, and then removed. Eggs from fowls kept on this plan show a large per cent. of fertility and yield vigorous chicks.

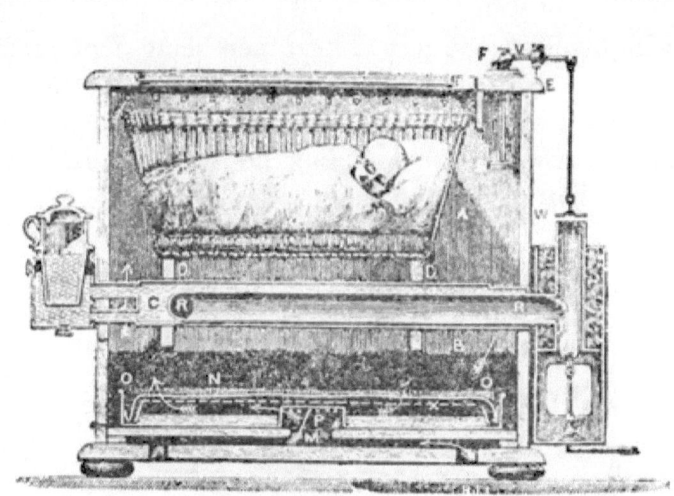

HEARSON'S AUTOMATIC NURSE.
For Nursing Weak or Premature Infants.

HATCHING DUCKS IN CHINA.

The artificial hatching of ducks is one of the interesting industries of China, and has been carried on extensively for hundreds of years. The Chinese are very fond of ducks and duck eggs, yet those in America are not far behind the Africo-Americans in their appetite for chicken. Most of the Chinese hatching houses are constructed of bamboo, plastered with mud and thatched with straw. The eggs are placed in baskets which have a tile bottom and a close straw cover. They are arranged around in rows and fire placed beneath. They are tested on the fifth day, and on the fifteenth day are placed on shelves and covered with blankets, the animal heat then being depended upon to finish the hatching. The natives along the coast who live on house boats or rafts get the eggs hatched at the hatching house and raise the ducks literally on the water.

CROCODILE EGGS.

Even if he does commit fo*u*l deeds on the Nile, the crocodile cannot be classed with poultry. Still it may interest many of our readers to know that it lays an egg smaller than a goose egg, the average size being three inches long and two inches in diameter, equally large at both ends, and very similar in shape to a snake egg. They are laid in the sand and hatched by the sun. On breaking the shell of an egg well advanced in incubation, you will find the young "croc" doubled up with his tail to his nose.

INDEX.

	PAGE
Incubation in Egypt,	5
Egyptian Incubating House,	8
A Good Incubator,	14
How to Choose an Incubator,	15
Don't Make a Failure,	17
The Best Size Incubator,	18
Hot Air or Hot Water?	19
Marking Eggs,	28
Table for Records,	32
Cooling the Eggs,	33
Testing Eggs,	35
How the Chicks Develop,	47
Animal Heat,	54
When Hatching,	54
Dead in the Shell,	55
Periods of Incubation,	63
Moisture in Hatching,	64
Hatching Ducks,	72
Hatching Geese,	73
Hatching Turkeys,	73
Hatching Ostriches,	78
A Letter,	79
The Thermostatic Incubator,	81
The Eureka Incubator,	84
The Eureka Brooder,	85
Improved Simplicity Hatcher,	86
Directions for Operating,	89
Simplicity Compartment Hatcher,	92

	PAGE
Water Expansion Regulators,	92
Two Regulators,	98
Hocus Pocus Regulators,	100
Other Methods of Regulation,	112
Thermostats,	115
Moisture Gauges and Hygrometers,	130
Brooding,	132
Brooding Houses,	137
Hot Air Brooding House,	138
Hot Water Brooding House,	140
Hot Water Stove with Apparatus,	143
Single Out-door Brooder,	144
Out-door Hot Water Brooder,	147
Brooder Yards,	148
Feeding Chicks,	148
Fattening Broilers,	151
Old Fowls and Young Chicks,	152
Selecting Breeding Stock,	153
Culling Breeding Stock,	155
When to Cull,	155
The Business Hen,	156
A Secret,	156
A Villainous Practice,	157
The Vaporizer and Its Use,	159
Egg and Broiler Farm,	162
Automatic Baby Nurse,	163
Hatching Ducks in China,	163
Crocodile Eggs,	164

www.ingramcontent.com/pod-product-compliance
Lightning Source LLC
Chambersburg PA
CBHW030252170426
43202CB00009B/712